Data Communication and Networking: A Practical Approach

MASSOUD MOUSSAVI, PH.D., P.E.
CALIFORNIA STATE POLYTECHNIC
UNIVERSITY-POMONA

DELMAR
CENGAGE Learning

Australia • Brazil • Japan • Korea • Mexico • Singapore • Spain • United Kingdom • United States

DELMAR
CENGAGE Learning™

Data Communication and Networking:
A Practical Approach
Massoud Moussavi

Vice President, Editorial: Dave Garza

Director of Learning Solutions: Sandy Clark

Acquisitions Editor: Stacy Masucci

Managing Editor: Larry Main

Senior Product Manager: John Fisher

Editorial Assistant: Andrea Timpano

Vice President, Marketing: Jennifer Baker

Marketing Director: Deborah Yarnell

Marketing Manager: Katie Hall

Marketing Coordinator: Jillian Borden

Production Director: Wendy Troeger

Production Manager: Mark Bernard

Content Project Manager: Barbara LeFleur

Production Technology Assistant:
  Emily Gross

Art Director: David Arsenault

Technology Project Manager: Joe Pliss

For product information and technology assistance, contact us at
Cengage Learning Customer & Sales Support, 1-800-354-9706
For permission to use material from this text or product,
submit all requests online at **www.cengage.com/permissions.**
Further permissions questions can be e-mailed to
permissionrequest@cengage.com

Library of Congress Control Number: 2011934129

ISBN-13: 978-1-111-12504-2

ISBN-10: 1-111-12504-X

Delmar
Executive Woods
5 Maxwell Drive
Clifton Park, NY 12065
USA

Cengage Learning is a leading provider of customized learning solutions with office locations around the globe, including Singapore, the United Kingdom, Australia, Mexico, Brazil, and Japan. Locate your local office at **www.cengage.com/global**

Cengage Learning products are represented in Canada by Nelson Education, Ltd.

To learn more about Delmar, visit **www.cengage.com/delmar**

Purchase any of our products at your local bookstore or at our preferred online store **www.cengagebrain.com**

Notice to the Reader
Publisher does not warrant or guarantee any of the products described herein or perform any independent analysis in connection with any of the product information contained herein. Publisher does not assume, and expressly disclaims, any obligation to obtain and include information other than that provided to it by the manufacturer. The reader is expressly warned to consider and adopt all safety precautions that might be indicated by the activities described herein and to avoid all potential hazards. By following the instructions contained herein, the reader willingly assumes all risks in connection with such instructions. The publisher makes no representations or warranties of any kind, including but not limited to, the warranties of fitness for particular purpose or merchantability, nor are any such representations implied with respect to the material set forth herein, and the publisher takes no responsibility with respect to such material. The publisher shall not be liable for any special, consequential, or exemplary damages resulting, in whole or in part, from the readers' use of, or reliance upon, this material.

Printed in the United States of America
2 3 4 5 6   19 18 17 16 15

Dedication

To my wife, Farkhondeh, and my daughters, Mehrnoush, Raha, and Ava who have continuously supported and helped me.

Contents

Preface

Data communication has helped us to be able to store, analyze, transmit, and receive very large volumes of information. This revolution in the transmission of information (which is a set of meaningful data) in the fastest possible time was made possible by technology that allows us to change analog signals (such as voice) to digital signals and vice versa. Networking also has made it possible for people around the world to communicate with each other in real time. As a result, data communication and networking have become the technology of the twenty-first century. To educate and train the workforce in this revolutionary technology, the fields of electronic and computer engineering, electronic and computer engineering technology, computer science, computer information science (CIS), and information technology (IT) have all included data communication and networking in their curricula. However, not all of these programs are interested in all aspects of data communication and networking. For example, while engineering or engineering technology programs are mostly focused on how to design, test, and build communication system hardware to improve and enhance the performance of current systems, computer science programs are mostly interested in the development and/or improvement of the software used in data communication and networking. The information technology programs are mostly interested in the design and management of network systems.

To respond to the needs of the above-mentioned programs for textbooks and reading materials, many excellent textbooks have been written and published by experts in this field, and I have used some of them in my years of teaching data communication and networking. My motivation for writing this textbook is very clear and simple. Most data communication and networking textbooks are written for CIS and IT or related fields. Most of them are very well-written textbooks that cover networking in depth and cover data communication only briefly. There is no single textbook that I'm aware of that presents both of these in detail or includes laboratory experiments, which are essential for electronic engineering or related programs. Consequently, I decided to write a data communication and networking textbook that explains all the possible methods of communication in this field, such as transmission of analog signals over digital media, transmission of digital signals over analog media, and conversion of data into signals. I wanted this textbook to also cover the current topics in the field of data communication and networking along with related laboratory experiments.

Each of the chapters on data communication includes related laboratory experiments to add hands-on experience to the theoretical knowledge that students gain from the chapter. All experiments are designed and tested via NI Circuit Design Suite/Multisim electronic simulation software. The setting value on each experiment must be adjusted when students are trying to construct and test it. Even more adjusting and precise settings are required when these experiments are to be built and tested in a hardware laboratory. Easy construction (of both software and hardware) and testing are the main reasons for selecting and designing these laboratory experiments.

Avoiding extensive writing and getting to the point of the subject as quickly as possible, as well as a clear and simple presentation, were my main focus in writing this textbook. I tried to explain each topic so that students would not need an understanding of high-level mathematics to make the subject easily understandable. The textbook should be suitable for all electronic and computer engineering and engineering technology students, as well as computer information science and information technology students at the undergraduate level.

As an educator, I hope that, by writing this textbook, I have taken one step toward enhancing students' knowledge in this fast-growing field that will evolve even further in the future.

SUPPLEMENTS TO THIS BOOK

Student Companion Site

A Student Companion Website is available containing MultiSIM files relating to Chapters 2, 3, 4, and 5.

ACCESSING A STUDENT COMPANION WEBSITE
SITE FROM CENGAGEBRAIN:

1. Go to: http://www.cengagebrain.com
2. Enter author, title or ISBN in the search window
3. When you arrive at the Product Page, click on the Access Now tab.
4. Click on the Student Resources link in the left navigation pane to access the resources

INSTRUCTOR SITE

An Instruction Companion Website containing supplementary material is available. This site contains an Instructor Guide, testbank, image gallery of text figures, and chapter presentations done in PowerPoint. Contact Delmar Cengage Learning or your local sales representative to obtain an instructor account.

ACCESSING AN INSTRUCTOR COMPANION WEBSITE SITE FROM SSO FRONT DOOR

5. Go to: http://login.cengage.com and login using the Instructor email address and password.
6. Enter author, title or ISBN in the **Add a title to your bookshelf** search box, click on **Search** button
7. Click **Add to My Bookshelf** to add Instructor Resources
8. At the Product page click on the **Instructor Companion site** link

NEW USERS

If you're new to Cengage.com and do not have a password, contact your sales representative.

ACKNOWLEDGMENTS

I would like to first thank my parents (Sharafat and Ismaiel) who have always encouraged their children to pursue their education and taught them to never stop learning. To them, education was the most important thing and I am fortunate that I inherited this belief from them.

I would also like to thank Ms. Stacy Massucci, the acquisitions editor, and Mr. John Fisher, the project manager at Delmar Cengage Learning, for their patience and continuous friendly and kind support throughout this project. Finally, I would like to thank Ms. Lina Rishmawi, at National Instrument who provided me a license to use the NI Circuit Design Suite/Multisim software, which I used to design and simulate the laboratory experiments in this textbook

I would like to express appreciation to the Ashley Green, Devil State Community College, Hamilton, Alabama, and Shiwen Mao, Auburn University, Auburn, Alabama, for their input as reviewers of this text.

1

Characteristics and Elements of a Communication System

Objectives

After completing this chapter, students should be able to:

- Describe the available transmission media and their characteristics.

- Explain signal transmission methods.

- Discuss communication impairments.

Communication is the transmission of a signal from one point that is called a *source* or *transmitter* to another point that is called a *destination* or *receiver*. Therefore, a communication system consists of three major parts: transmitter (source), transmission line (transmission medium), and receiver (destination). The input signal at the source can be a continuous (analog) signal or a discrete (digital) signal. In data communication, a digital signal is a symbolic representation of information. The ultimate goal of a communication system is to transmit a signal from a source to a destination without distortion and noise, and with minimal time delay. To transmit a signal effectively and efficiently, the transmitted signal needs to be modulated before transmission. There are many different methods of modulation, and a signal can be modulated based on its characteristics. For example, an analog signal has three characteristics (parameters): amplitude, frequency, and phase. Therefore, to transmit an analog signal, it needs to be modulated either by amplitude modulation (AM), frequency modulation (FM), or phase modulation (PM) techniques. Analog modulation methods are not covered in this book, but conversion or modulation of analog to digital signals, digital to analog signals, and digital data to digital signals will be covered in Chapters 2, 3, and 4.

1.1   COMMUNICATION SYSTEM

Communication Elements

Communication elements consist of a transmitter, a transmission line, and a receiver. The transmitter is an electronic device that has the ability to modulate or encode information in an analog or digital format and send it to a receiver over a transmission medium. The transmission line is a medium (either guided, such as twisted wire, coaxial cable, or optical fiber; or unguided, such as air) to transfer the modulated or encoded output signal from a transmitter to a receiver with minimal loss of signal strength. The degree of signal loss in the transmission line depends on its physical and electronic characteristics. The receiver is an electronic device that has the ability to receive a modulated or encoded signal from a transmission line and then demodulate or decode it.

Information

Information is a collection of data in one of two forms: analog (such as voice, power, pressure, heat, or light) or discrete (such as an analog signal that is sampled and quantized). An analog signal is an electromagnetic wave and is represented by a series of sine waves,

$$E(t) = E_0 \sin (\omega t + \varphi) \tag{1.1}$$

where E_0 is the amplitude of the signal at time zero, ω is angular frequency, $\omega = 2\pi f$, and φ is the phase of the signal. The main parameters of an analog signal are amplitude, frequency, and phase. Figure 1–1 shows a time-varying analog signal.

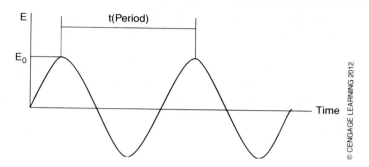

© CENGAGE LEARNING 2012.

Figure 1–1: Time-varying analog signal

Example 1: A time-varying analog signal transmits through a coaxial cable according to the following equation: $E(t) = 15$ V sin $[(120\pi) \, t + \pi/6]$. Find its amplitude, frequency, and time for one complete cycle (period) and phase of the signal. At what time does the signal have maximum strength?

Solution: Comparing the signal's equation and Equation (1.1), we have:

$$E_0 \text{ (amplitude)} = 15 \text{ V}$$
$$120 \, \pi = 2\pi f$$
$$f = 120/2 = 60 \text{ Hz}$$

$$t \text{ (time for one cycle or period)} = 1/f = 1/60 = 16.67 \text{ m sec}$$
$$\varphi = \pi/6 \text{ (radiant)} = 30 \text{ degree}$$

The maximum strength of the signal happens when the signal value is equal to its amplitude. That means:

$$E = E_0$$
$$\sin [(120\pi) \, t + \pi/6] = 1$$
$$[(120\pi) \, t + \pi/6] = \pi/2$$
$$(120\pi) \, t = -\pi/6 + \pi/2 = \pi/3$$
$$t = (\pi/3)/(120\pi)$$
$$t = 2.77 \text{ m sec}$$

The data that is going to be transmitted over a digital communication system is represented in two voltage levels: positive and negative, or positive and zero (ground). It is customary to show the positive voltage level by logic 1 (bit 1) and the zero voltage level by logic 0 (bit 0). In data communication, the transmitted information consists of many symbols and/or characters that are represented by series of bit 1 and bit 0. These symbols or characters are usually coded in the ASCII or Unicode formats.

Example 2: Convert the letter "C" to ASCII (7-bit), extended ASCII (8-bit), and Unicode formats.

Solution: ASCII stands for American Standard Code for Information Interchange and is a 7-bit code that represents 128 characters. The extended ASCII code is an 8-bit code based on the ISO 8859-1 and Microsoft Windows Latin-1 standards. Unicode (Universal Code) is a multiblock of 16-bit code that represents the characters of all languages where each block belongs to one specific language.

The letter "C" is represented by "43 or 1000011" in the ASCII code.

The letter "C" is represented by "043 or 01000011" in the extended ASCII code.

The letter "C" is represented by "U0043" in Unicode.

Transmission Lines or Transmission Media

Transmission lines are divided into two categories: guided or bounded media and unguided or unbounded media. In a guided medium the transmitted signal will be guided through or bounded within a line or channel for transmission. In an unguided or unbounded medium the transmitted signal is not guided by any type of cable for transmission. Air is an unguided medium that is used for wireless transmission. An interface cable (such as Small Computer Systems Interface [SCSI] and FireWire) or interface bus (such as a Universal Serial Bus [USB]) is used for data transmission between computers and other devices, such as a CD drive, DVD drive, hard drive, and scanner. Basically, these transmission media are either hardware or software regulators that control data transmission between two or more devices. SCSI, USB, and FireWire (IEEE 1394) are the most common interfaces in data exchange.

Shielded and Unshielded Twisted Pair Wire

A twisted pair wire consists of two wires: one that carries a positive signal and another that carries a negative signal. To distinguish the opposite polarity nature of these two wires, they are shown by gray and black in Figure 1–2. The advantage of having two wires is that the 180-degree phase difference between the two wires cancels out noise. There are two types of twisted pair wires: unshielded (UTP), as shown in Figure 1–2, and shielded (STP), as shown in Figure 1–3. Shielded twisted pair wire is protected from noise interference and therefore has better noise reduction characteristics than an unshielded twisted pair. STP cable is mainly used in the IBM networking system. The Electronic Industries Association (EIA) divides UTP cable into eight categories. CAT-1 with low data rates (less than 2 Mbps), are used in telephone wiring, including T-1 lines. CAT-2, CAT-3, and CAT-4, with data rates of 10, 16, and 100 Mbps, respectively, are used in local area networks (LANs). CAT-5E, CAT-6, and CAT-7 are new categories with higher data rates—125, 200, and

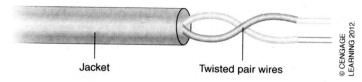

Jacket   Twisted pair wires

© CENGAGE LEARNING 2012.

Figure 1–2: Unshielded twisted pair (UTP) wires

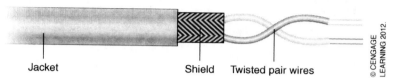

Jacket   Shield   Twisted pair wires

© CENGAGE LEARNING 2012.

Figure 1–3: Shielded twisted pair (STP) wires

600 Mbps, respectively, and they are also used in local area networking. Individually shielding pairs of wires with a helical metallic foil in addition to the outer shielding has enabled CAT-7 cable to transfer data at a much higher data rate with decreased effect of crosstalk. The 8-pin RJ-45 and 4-pin RJ-11 (RJ stands for Registered Jack) connectors are the most common connectors for connecting UTP cables.

Coaxial Cable

Coaxial cable was developed to transmit high-speed analog signals (about 10,000 voice signals) and high-rate digital signals (up to 600 Mbps) for radio frequency in long-distance communication. As shown in Figure 1–4, a coaxial cable consists of a solid metallic core (specifically copper) wrapped in a plastic insulator (inner insulator), a tubular metallic shield, and an outer plastic insulator. The disadvantage of coaxial cable over twisted pair cable is in signal attenuation. Signal attenuation in coaxial cable increases nonlinearly as the frequency of the signal increases.

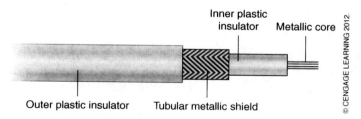

Inner plastic
insulator   Metallic core

© CENGAGE LEARNING 2012.

Outer plastic insulator   Tubular metallic shield

Figure 1–4: Structure of a coaxial cable

There are several varieties of coaxial cable; these are mostly divided into three main categories based on core diameter (impedance), size, and insulation material. These categories are identified by Radio Guide (RG) numbers. The impedance value and application of coaxial cable types are listed in Table 1–1.

Type	Impedance (Ω)	Application
RG-11	50	Thick Ethernet (10Base5)
RG-58	50	Thin Ethernet (10Base2)
RG-59	75	Cable TV

© CENGAGE LEARNING 2012.

Table 1–1: Coaxial cable types

RG-6 coaxial cable is a larger and cheaper version of RG-59 cable; RG-6 coaxial cables have an 18-AWG copper-coated steel center conductor and have higher bandwidth than RG-59 coaxial cables.

BNC and BNC-T connectors (see Figure 1–5) are the most common connectors used to connect coaxial cables.

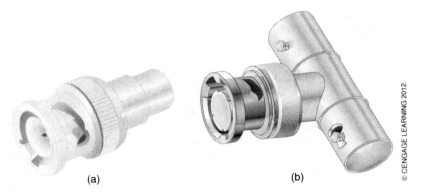

(a)   (b)

© CENGAGE LEARNING 2012.

Figure 1–5: BNC connectors: (a) BNC male to RCA female (b) BNC, T-connector (one male, two females) (www.apexcctv.com)

RS232 (RS232C)

The RS232 standard or EIA 232 is an interface device that is designed to connect data terminal equipment or DTE (such as a computer) that is equipped with a male DB25 connector to data circuit-terminating equipment or DCE (such as a modem) that is equipped with a female DB25 connector for data transmission. The DB25 is used mostly for serial connections, allowing for asynchronous data transmission. Personal computers are equipped with a smaller version of the DB25 connector known as a DB9 connector. Figure 1–6 shows both female DB25 and DB9 connectors. The pin assignment for DB9 and DB25 connectors and their conversion are shown in Tables 1–2, 1–3, and 1–4.

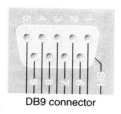

DB9 connector   DB25 connector

Figure 1–6: DB9 and DB25 female connectors

Pin assignment for DB9 connector			
1. Data carrier detect	2. Receive data	3. Transmit data	4. Data transmit ready
5. Signal ground	6. Data set ready	7. Request to send	8. Clear to send
9. Ring indicator			

Table 1–2: Pin assignment for DB9 connectors

Pin assignment for DB25 connector			
1. Protective ground	2. Transmit data	3. Receive data	4. Request to send
5. Clear to send	6. Data send ready	7. Signal ground	8. Data carrier detect
9. Test pin	10. Test pin	11. Unused	12. Data carrier detect-2
13. Clear to send (2)	14. Transmit data (2)	15. Transmitter clock (DCE)	16. Receive data (2)
17. Receiver clock	18. Local loopback	19. Request to send-2	20. Data terminal ready
21. Signal quality detector	22. Ring indicator	23. Data signal rate detector	24. Transmitter clock (DTE)
25. Test mode			

Table 1–3: Pin assignment for DB25 connector

DB9	DB25	Function
1	8	Data carrier detect
2	3	Receive data
3	2	Transmit data
4	20	Data terminal ready
5	7	Signal ground
6	6	Data set ready
7	4	Request to send
8	5	Clear to send
9	22	Ring indicator

Table 1–4: DB9 to DB25 conversion

RS232 devices are designed for asynchronous (transmission at any time) and serial (bit-by-bit transmission) communication. In asynchronous transmission, the receiver is able to begin receiving data at any time, including at an incorrect time. To overcome this problem, two start and stop bits are added to the data being transmitted to notify the receiver of the start and stop time of the data transmission. The data frame format for RS232 is as follows: Start bit (bit "0," an OFF or Space state), Data bits, Parity bit for error detection purposes, and the Stop bit (bit "1," an ON or Mark state). RS232 is designed as a 50-foot long cable with a data transmission bit rate of 19.2 Kbps. RS232 is also available with longer cable lengths and much lower transmission bit rates; for example, the 1000-foot cable has a transmission bit rate of only 4.8 Kbps. There are two positive and negative voltage levels to represent the signal of the RS232 with the following values:

OFF/Space state: (+5 V to +15 V at the transmitter side)
and (+3 V to +25 V at the receiver side).

ON/Mark state: (−5 V to −15 V at the transmitter side)
and (−3 V to −25 V at the receiver side).

The conductors of the RS232 are called *circuits*. The circuits are divided into five groups: ground (three lines), data (two lines), control (eight lines), timing (three lines), and secondary channel (five lines). The eight control lines of the RS232 are as follows: data terminal ready (DTR), data carrier detect (DCT), data set ready (DSR), ring indicator (RI), request to send (RTS), clear to send (CTS), transmitted data (TxD), and received data (RxD).

Optical Fiber or Optical Waveguide

Optical fiber is a transmission medium that carries optical signals. Optical fiber consists of a core and cladding as shown in Figure 1–7. The core is a very tiny fiber (with a diameter of about 8 to 200 micrometers depending on the type of fiber) of glass or plastic that carries the light signal and the cladding provides the necessary total internal reflection between the core and cladding boundary. The diameter range of cladding is between 125 and 140 micrometers.

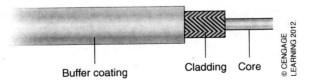

Buffer coating   Cladding   Core

© CENGAGE LEARNING 2012.

Figure 1–7: Core and cladding in an optical fiber

Fiber optic cable has many advantages over coaxial and twisted pair cables. The following are the main advantages of fiber optics:

- Almost unlimited bandwidth
- Smaller size and weight
- Immunity from electromagnetic and radio frequency interferences
- Lower signal attenuation
- Immunity from crosstalk
- Resistance to electric spark, heat, radiation, and corrosion
- Lower bit rate error
- Better security (difficult to tap into)

The main disadvantages of fiber optics lie in the technical difficulties of connecting optical fibers. While an electrical signal is the only signal type that is used in a copper wire such as twisted pair, when using fiber optic cable an electrical signal must first be converted to light at the transmitter and then converted from light back to an electrical signal at the receiver.

There are three basic optical fiber types: single mode, multimode-step index, and multimode-graded index. Single-mode optical fiber transmits in only one path or transverse mode. Because of the very small core diameter (2 to 8 micrometers) and lower index of refraction, the input light signal enters and leaves the fiber in only one path with negligible delay. For the same reason, the acceptance angle (the angle that gathers the most incident light rays) and numerical aperture (the measure of the fiber's ability to gather incident light rays) are both very small; this makes launching an incident light into this type of fiber very difficult. Lasers are a suitable light source for single-mode fiber because lasers emit a coherent light. Single-mode fiber has a very large bandwidth compared to multimode fibers. The refractive index profile and the pattern of light travel in single-mode fiber are shown in Figure 1–8.

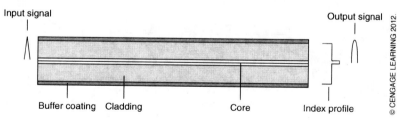

Figure 1–8: Single-mode fiber

Multimode-step index fiber has a much larger core diameter (ranging from 50 to 200 micrometers) than single-mode fiber. There are two different types of light rays in multimode-step index fiber: axial and marginal rays. Axial rays travel along the central axis of the core. Marginal rays that enter the fiber at an angle will reflect back to the core when they reach the boundary between the core and cladding, since the index of refraction of the core is slightly higher than the index of refraction of the cladding. Compared to axial rays, the marginal rays travel a greater distance in the fiber, and therefore take more time to exit the fiber. This causes dispersion in the light signal, which is called *modal dispersion*. The modal dispersion in the multimode-step index fiber lowers the bit rate and bandwidth in comparison with single-mode fiber. The refractive index profile and the pattern of light travel in multimode-step index fiber are shown in Figure 1–9.

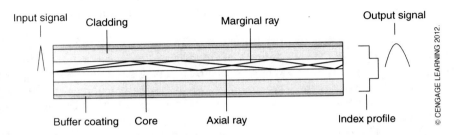

Figure 1–9: Multimode-step index fiber

Multimode-graded index fiber has a core with an index of refraction that decreases gradually as it reaches the cladding; this causes light rays to travel in a helical pattern in a much shorter time compared to multimode-step index fiber. As a result, modal dispersion in this type of multimode fiber will decrease significantly and consequently leads to a much higher bit rate and bandwidth compared with multimode-step index fiber. The refractive index profile and the pattern of light travel in multimode-graded index fiber are shown in Figure 1–10.

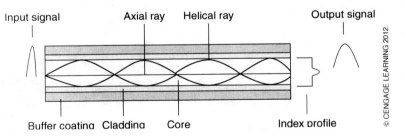

Figure 1–10: Multimode-graded index fiber

There are several varieties of connectors that can be used to connect two optical fibers with different levels of performance and fiber types. For example, the Fixed Connection (FC) is the most popular connector for single-mode optical fiber but needs special attention for proper alignment. Signal loss is the major issue in selecting connectors. Most of the losses are due to misalignments when two optical fibers are connected together. Figure 1–11 shows three different optical fiber connectors.

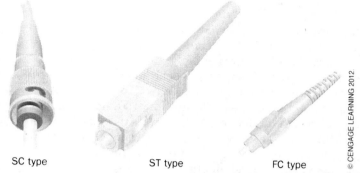

Figure 1–11: Three major types of optical fiber connectors: Snap-in Connector (SC), straight-tip (ST), and FC (www.fiberoptics4sale.com)

Small Computer Systems Interface

Small Computer Systems Interface (SCSI) was developed primarily as a parallel interface for hard disk drives. It is now used as a high-performance and high-data-rate peripheral interface for connecting or daisy-chaining multiple input/output devices such as hard disk or CD-ROM drives. In fact, SCSI is a system-level bus that comes with different standards, cable types, and connector types. Figure 1–12 shows some of the SCSI connector types. SCSI usually is connected to a controller card in order to send and receive data. It is also widely used in RAID (Redundant Array of Independent Discs) technology.

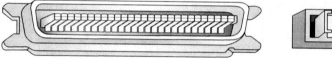

Figure 1–12: Centronics 50 and 68 pin connectors (www.belkin.com)

The original SCSI (SCSI-1 or narrow SCSI) was an 8-bit bus width with a maximum data transfer rate of 5 Mbps through a cable with a 50-pin connector. Table 1–5 shows specifications for different types of SCSI.

SCSI types/ specifications	Data rate (Mbps)	Bus width (bits)	Maximum number of supported devices	Connector type (pins)
SCSI-1	5	8	8	50
Ultra SCSI	20	8	4-8	50
Wide Ultra SCSI	40	16	4-8	68
Wide Ultra SCSI-2	80	16	16	68
Ultra-320 SCS-3	320	16	16	68

Table 1–5: Different types of SCSI and their specifications

Universal Serial Bus

The Universal Serial Bus (USB) is a serial bus that can connect up to 127 devices that require a high-speed connection, such as printers, joysticks, and digital cameras, to a computer. USB is a plug-and-play and hot-pluggable device with a maximum data rate of 480 Mbps. It supports isochronous devices. Both Windows and Macintosh operating systems will recognize any USB device as soon as it is connected to the computer. Most computers have a limited number of USB ports but it is possible to connect many USB devices to a computer through a USB hub (hubs will be discussed in Chapter 10). There two types of USB connectors: the A type, which has a rectangular shape and is connected to the computer; and the B type, which has a square shape and can be found on other devices that need to be connected to the computer. Figure 1–13 shows both types of USB connectors.

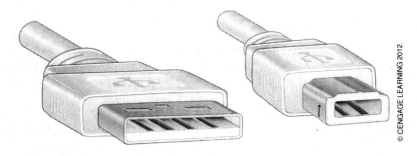

© CENGAGE LEARNING 2012.

Figure 1–13: A type (left) and B type (right) USB connectors

FireWire or IEEE-1394

IEEE-1394 is the standardized version of FireWire, which was originally developed by Apple in 1995 to transfer data to and from high-bandwidth audio and video digital devices, such as a digital camera or camcorder. FireWire is also called *i.Link*. FireWire is very similar to USB (both are serial buses, plug-and-play, hot-plug, and support isochronous devices) but much faster (up to 800 Mbps compared with USBs, which have a maximum rate of 480 Mbps) and able to connect up to 63 devices in a daisy-chain connection. This means FireWire is able to connect two digital devices to exchange data without connecting them to a computer (peer-to-peer networking, which will be discussed in Chapter 6). There are two versions of FireWire, the original version or FireWire-400 and the new version or FireWire-800, with data transfer rates of 400 Mbps and 800 Mbps, respectively. Both Windows (98 or later versions) and Macintosh operating systems support FireWire; therefore, any computer with these operating systems will recognize and communicate with a FireWire device as soon as it is connected to the computer. FireWire-400 and FireWire-800 are attached to a computer through 6-pin and 9-pin connectors, respectively, as shown in Figure 1–14.

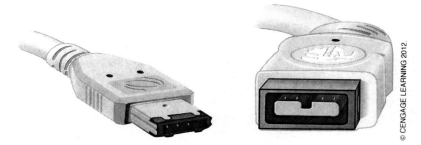

© CENGAGE LEARNING 2012.

Figure 1–14: 6-pin and 9-pin FireWire connectors

FireWire uses the IEEE-1212 standard for its 64 bits, addressing criteria for storing information, recognizing the transmitter device, and transmitting data, with 48 bits for storage area, 10 bits for bus ID, and 6 bits for physical ID.

1.2 SIGNAL TRANSMISSION METHODS

A signal can be transmitted over a transmission medium using two different methods: baseband and broadband.

Baseband Transmission Method

The baseband transmission method is used for short-distance transmission where the entire bandwidth of the medium is used to carry only one signal. Therefore, only one message or piece of information is transmitted at a time. If there is a need for sending multiple messages at a time, interleaving is necessary; this is possible using time-division multiplexing. Baseband transmission is used in Ethernet and Token Ring networks.

Broadband Transmission Method

Broadband transmission is used to transmit multiple signals from multiple channels at the same time over at the same time one single medium. Broadband transmission is capable of transmitting text, audio, and video over long distances. A good example of broadband transmission is cable TV, which carries multiple channels for subscribers using frequency-division multiplexing. Multiplexing methods will be covered in Chapter 5.

1.3 ATTENUATION, NOISE, AND DISTORTION

Signals can be corrupted during transmission and this is called *transmission impairment*. Transmission impairments occur as a result of attenuation, noise, and distortion.

Attenuation

Attenuation is the loss of strength of a signal as it travels a distance. Attenuation varies as a function of frequency. Figure 1–15 shows the effects of attenuation on a signal as it travels.

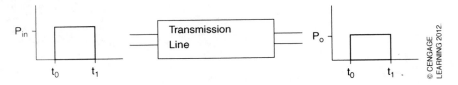

Figure 1–15: Signal attenuation in a transmission line $P_0 < P_{in}$

Noise

Noise is a random variation in an electrical signal. Communication systems are designed and manufactured on either passive or active electronic components. Both types of components are made of either conductive or semiconductive materials. Applied voltage will boost the number of free electrons in these materials. These free electrons move randomly, fluctuate, and collide with the molecule of the conductor or semiconductor and as a result they will generate noise. In 1918, Walter Schottky published his findings on the impact of the random motion or fluctuation of free electrons on communication. Schottky stated that there are two types of fluctuation noise in electronic circuits as a result of random motion or fluctuation of charge: free electrons (thermal noise) and random emission of charge, that is, fluctuation in the electrical current (shot noise). Figure 1–16 shows the shape of a signal with and without the effect of noise. Thermal noise and shot noise are the two major sources of noise in digital communication.

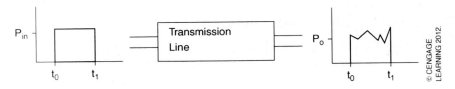

Figure 1–16: The effect of noise on a signal during transmission

Thermal Noise

Voltage applied to a conductor or semiconductor device will increase its temperature, which causes free electrons to move more freely and as a result generate noise, called *thermal* or *Johnson noise*. In 1920, J.B. Johnson, after extensive study of shot noise, concluded that thermal noise is a more fundamental type of noise than shot noise when an electrical charge fluctuates. Thermal noise is directly proportional to the operating bandwidth (BW) and absolute temperature (T in Kelvin). The proportional factor is called the *Boltzmann's constant, k* (1.38×10^{-23} J/K)

$$P_n = k \cdot T \cdot BW \qquad (1.2)$$

Example 3: A digital communication system operates at a bandwidth of 6 MHz and temperature of 400 K. Determine the thermal noise power.

Solution:

$$P_n = (1.38 \times 10^{-23} \text{ J/K})(400 \text{ K})(6 \times 10^6 \text{ Hz})$$
$$P_n = 3.312 \times 10^{-14} \text{ W}$$

Example 4: Determine the noise voltage of the digital communication system of Example 3 if the load impedance is 1.56 MΩ.

Solution: Electrical power and voltage are related to each other according to the following equation:

$$P = \frac{V^2}{Z},$$

therefore, the voltage noise is equal to:

$$V_n = \sqrt{k \cdot T \cdot BW \cdot Z} \qquad (1.3)$$

$$V_n = \sqrt{(1.38) \times 10 - 23 \,(400)\,(6 \times 106)\,(1.56 \times 106)}$$

$$V_n = 2.27 \times 10^{-4} \text{ V}$$

The noise temperature, T, can be determined easily if we know either the noise power or noise voltage from Equations (1.2) and (1.3). Keep in mind that the bandwidth and impedance are characteristics of a communication system.

Shot Noise

Shot noise is the random fluctuation of the electric current. While in a conductor, shot noise is the result of random movement or fluctuation of free electrons, in a semiconductor, shot noise is the result of electron-hole recombination and minority diffusion processes. In shot noise, the electric current fluctuation, σ, follows the Poisson distribution and has a standard deviation of:

$$\sigma = \sqrt{2 \cdot I \cdot q \cdot BW} \qquad (1.4)$$

where I is the saturation current in ampere and q is an electron charge (1.59×10^{-19} C).

When this shot noise current flows through, load impedance results in a shot noise power of:

$$P_n = 2 \cdot I \cdot q \cdot BW \cdot Z \tag{1.5}$$

Example 5: Find the minimum shot noise power in a digital communication system when a 100-mA current flows through a 1-MΩ load impedance.

Solution: The minimum noise power is due to the minimum bandwidth, that is, 1 Hz.

$$P_n = 2 \, (100 \times 10^{-3} \, \text{A}) \, (1.59 \times 10^{-19} \, \text{C}) \, (1 \, \text{Hz}) \, (1 \times 10^{6})$$
$$P_n = 3.18 \times 10^{-14}$$

Signal-to-Noise Ratio

Noise in a communication system will disturb its performance tremendously. The signal-to-noise ratio (SNR) is the most important factor in determining the performance of a receiver. The relation between signal and noise at the receiver expressed in decibels (dB) is given by the following equation.

$$\text{SNR (dB)} = 10 \log \frac{p_s}{p_n} \tag{1.6}$$

where p_s and p_n are the signal strength and noise strength, respectively, at a receiver.

Example 6: Determine the signal-to-noise ratio at the input of a receiver in decibels if the signal strength is 25 nW and the noise strength is 100 pW.

Solution:

$$\text{SNR (dB)} = 10 \log \frac{p_s}{p_n} = 10 \log \frac{25 \, \text{nW}}{100 \, \text{pW}} \quad 10 \log (0.25 \times 10^{3})$$

$$\text{SNR (dB)} = 10 \, (2.398) = 23.98$$

Example 7: Determine the signal strength at the input of a receiver if the noise strength is 80 pW and the SNR (dB) is 20.97.

Solution:

$$\text{SNR (dB)} = 10 \log \frac{p_s}{p_n}$$

$$p_s = p_n \log^{-1} \frac{\text{SNR (dB)}}{10}$$

$$p_s = (80 \text{ pW}) \log^{-1} \frac{20.97}{10}$$

$$p_s = (80 \text{ pW}) (125) = 10000 \text{ pW} = 10 \text{ nW}$$

Distortion or Delay Distortion

The propagation velocity of a signal depends on the frequency. In a guided medium, the various frequency components of a signal will arrive at the receiver at different times, which causes signal deformation. This is also called *distortion*. Distortion or delay in digital data will cause part of one bit to move into the position of the next bit, which is called *intersymbol interference* (ISI). Figure 1–17 shows distortion of a signal as a result of propagation delay.

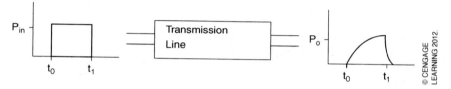

Figure 1–17: Delay in a transmitted signal

1.4   CHANNEL CAPACITY

In data communication, the ultimate goal is to send information as fast as possible without data corruption and error. To reach this goal, we are faced with some dilemmas such as bandwidth of the channel and noise in the channel. A channel is either noise-free/noiseless or noisy. In 1927, H. Nyquist proposed that the pulse rate should be equal to or less than twice the bandwidth:

$$f_p \le 2BW \tag{1.7}$$

Later, in 1928, Hartley, through his law, stated that the maximum number of distinct pulses (M) that can be transmitted over a channel is limited by the dynamic range of the signal amplitude and ability of the receiver to distinguish these amplitude levels. Therefore, according to Hartley's law, the maximum information (data) rate is equal to:

$$f_b = f_p \log_2 M \tag{1.8}$$

To find the maximum information (data) rate in a noiseless or noise-free channel, we need to substitute Equation (1.7) into Equation (1.8):

$$f_b \le 2BW \cdot \log_2 M \tag{1.9}$$

Example 8: Calculate the maximum data rate (bit rate) for a noise-free telephone line that transmits a two-level signal.

Solution: The appropriate bandwidth for a telephone line is about 4 kHz. Therefore, based on Nyquist's theorem, the maximum data rate is:

$$f_b = 2 \ (4 \text{ kHz}) \ (\log_2 2) = 8 \text{ Kbps}$$

Example 9: Repeat Example 8 for transmitting a three-level signal.

Solution: We may convert a base-2 log to a base-10 log as follows:

$$\log_2 M = (\log_{10} M)/(\log_{10} 2) = 3.322 \ (\log_{10} M) \tag{1.10}$$

$$f_b = 2 \ (4 \text{ kHz}) \ (3.322 \log_{10} 3) = 8 \ (3.322) \ (0.477) = 12.68 \text{ Kbps}$$

Example 10: How many signal levels are required to transmit information over a noise-free channel with a bandwidth and data rate of 10 kHz and 40 Kbps?

Solution:

$$40 \text{ Kbps} = 2 \ (10 \text{ kHz}) \log_2 M = 2 \ (10 \text{ kHz}) \ (3.322 \log_{10} M)$$
$$M = \log^{-1}(40/20(3.322) = 4 \text{ signal levels}$$

In 1948, C.E. Shannon proposed a theorem on channel capacity partly based on Nyquist and Hartley's work on information rate. In his theorem, Shannon stated that the maximum data rate that can be sent over a channel subject to additive white Gaussian noise (noisy channel) is equal to:

$$C = BW \cdot \log_2 (1 + S/N) \tag{1.11}$$

The channel capacity in a base-10 log is:

$$C = 3.322 \ BW \cdot \log_{10} (1 + S/N) \tag{1.12}$$

Example 11: What is the channel capacity of a telephone line if the signal-to-noise ratio is 25 dB?

Solution:

$$S/N = 25 \text{ dB} = \log^{-1} (25/10) = 316.23$$
$$C = 3.322 \ (4 \text{ kHz}) \log_{10} (1 + S/N) = 13.288 \log_{10} (1 + 316.23)$$
$$C = 33.23 \text{ Kbps}$$

SUMMARY

In a communication system, an electrical signal travels from one point (transmitter) to another point (receiver) through either a guided medium or an unguided medium. A guided medium is a bounded physical path that the signal travels through. Examples of guided media are conductive twisted and untwisted pairs of wires, coaxial cables, and optical fiber. An unguided medium such as air has no boundary and the signal is free to travel in any direction. Wireless communication uses an unguided medium. Some guided media, such as Small Computer Systems Interface (SCSI), Universal Serial Bus (USB), and FireWire, have been designed and developed for data transmission from one computer to another one. These types of guided media are in the category of bus technology. Each type of medium has advantages and disadvantages over the others; nevertheless, each has specific applications for which they are best suited.

Signals are either natural (analog) or manmade (digital). An analog signal cannot travel through a digital medium, and a digital signal cannot travel through an analog medium without appropriate modulation and demodulation.

As a signal travels through a medium, it will experience loss of power (attenuation), change in its original shape (distortion), or be affected by internal and external noise. Thermal noise and shot noise are the most common sources of noise in data communication. To recover the original signal at the receiver, the ratio of signal and noise strength should be limited. Therefore, the signal-to-noise ratio, which is measured in decibels, is an important factor in any communication.

The ultimate goal in data communication is to transmit a large volume of information as fast as possible. There are two major factors that have an effect on reaching this goal: bandwidth and noise. In 1948, Shannon proposed a theorem on calculating the data rate in a medium or channel partly based on the work of Nyquist and Hartley. The Shannon channel capacity theorem has helped us to determine the data rate based on a given bandwidth and signal-to-noise ratio.

Review Questions

Questions

1. What are the advantages of UTP over STP?

2. What are the advantages of coaxial cable over twisted pair cables?

3. Why must the index of refraction of cladding be lower than the index of refraction of the core in an optical fiber?

4. What is the difference between a single-mode and multimode optical fiber?

5. What is the frequency range of visible light?

6. What are the advantages and disadvantages of using optical fiber over coaxial or twisted pair cables?

7. Which transmission line has a larger bandwidth?

8. Which transmission line can transmit signal for a longer distance?

9. Which transmission line has better bit rate value, a 2-level signal or 4-level signal? Why?

10. What would be the signal-to-noise ratio in terms of input and output voltages?

11. Why are there Start and Stop bits when an RS232 cable is used to communicate between two devices?

12. What is the maximum data transfer rate for RS232?

13. List the five groups into which the RS232 circuit is divided.

14. Name and describe all eight control lines of RS232.

15. What are the voltage levels of RS232?

16. What are the major differences between DB9 and DB25 connectors?

17. How does SCSI connect multiple devices together?

18. What is the range of data transfer rate in SCSI?

19. How many devices can be added together by SCSI?

20. How many devices can be connected to a computer by USB?

21. How is it possible to add more devices to computer with a limited number of USB ports?

Review Questions: continued

22. List the major characteristics of a FireWire device.

23. How does a FireWire device get its power?

24. What are the differences between USB and FireWire?

Problems

1. A time-varying analog signal transmits through a coaxial cable according to the following equation: $E(t) = 12V \sin [(90\pi) t + \pi/3]$. Find its amplitude, frequency, and time for one complete cycle (period), and phase of the signal. At what time is the signal at its maximum strength?

2. Represent the word "Data" in ASCII, extended ASCII, and Unicode.

3. Determine the noise voltage of the digital communication system in Example 3 if the load impedance is 2.2 MΩ.

4. Determine the signal strength at the input of a receiver if the noise strength is 40 V and the SNR (dB) is 21.

5. Find the minimum shot noise power in a digital communication system when a current of 150 mA flows through a 2.2-MΩ load impedance.

6. Calculate the maximum data rate (bit rate) for a noise-free transmission line that transmits an eight-level signal.

7. Compute the shot noise power spectral density and noise voltage of a semiconductor device that is used in a digital communication system with a saturation current of 10 mA.

8. Determine the noise power and noise voltage of a digital circuit operating at a bandwidth of 5 MHz and temperature of 420 K with a load impedance of 1 MΩ. Assume an ideal noise source.

9. A digital communication system generates a noise voltage of 30 μW. If the noise temperature is 350 K and the load impedance is 750 Ω, what would be the operating bandwidth?

10. Calculate the noise power of a digital communication receiver operating with a bandwidth of 36 MHz and the equivalent noise temperature of 120 K.

11. Calculate the signal-to-noise ratio in dB at the input of communication equipment with a signal strength of 20 nW and a noise level of 100 pW.

Continues on next page

12. What is the channel capacity of a telephone line if the signal-to-noise ratio is 16 dB?

13. How many signal levels are required to transmit information over a noise-free channel with bandwidth of 25 kHz and data rate of 100 Kbps?

14. Calculate the signal power at the input of an attenuator with a noise figure of 3 dB and an output signal of 10 nW.

Connecting a 4-pair unshielded twisted pair (UTP) cable to an RJ-45 connector.

Materials for this activity:

> One 4-pair UTP cable
>
> One RJ-45 connector
>
> One wire cutter
>
> One wire stripper

A 4-pair UTP cable consists of the following pairs:

> Pair 1: Blue and white/blue
>
> Pair 2: Orange and white/orange
>
> Pair 3: Green and white/green
>
> Pair 4: Brown and white/brown

The RJ-45 connector has eight connector slots. Figure 1–18 shows the connector slot assignments for the 4-pair wires of the UTP cable.

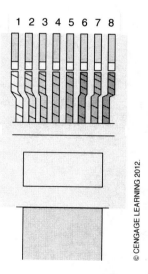

© CENGAGE LEARNING 2012.

Figure 1–18: The RJ-45 connectors and the connector slot assignments

Continues on next page

The connector slot assignments are as follows:

Connector slots 4 and 5 are assigned to pair 1.

Connector slots 1 and 2 are assigned to pair 2.

Connector slots 3 and 6 are assigned to pair 3.

Connector slots 7 and 8 are assigned to pair 4.

Procedure:

Use the wire cutter to cut the UTP cable to a desired size and make sure all wires are cut evenly. Use the wire stripper to remove the plastic insulator from the conductor wire for about a quarter of an inch.

Hold the RJ-45 as shown in Figure 1–18.

Insert the proper pair of wires into the assigned connector slots according to Figure 1–18.

Use the crimping tool to secure the wires in the connector slots. To make sure you have a secure connection, try to gently pull the UTP cable from the RJ-45. Wires may come loose if they are not connected securely.

2

Signal Conversion, Case One: Analog-to-Digital Signal Conversion

Objectives

After completing this chapter, students should be able to:

- *Discuss the conversion processes of an analog signal to a digital signal.*

- *Introduce uniform and nonuniform quantization.*

- *Describe different pulse code modulation techniques.*

- *Analyze errors in the quantization process.*

The most common data that we transmit over a digital communication system is an analog (continuous) signal, such as voices. These types of data cannot be transmitted unless they are converted to a digital (discrete) signal in order to be recognizable by digital communication machines. Digitization, quantization, and encoding are the processes that convert an analog signal to a digital signal for transmission. Pulse code modulation (PCM) is the most common technique that is used for this conversion.

2.1 DIGITIZATION

Digitization is a process that converts a continuous signal to a discrete signal, which is also called a *sampling process*. Pulse amplitude modulation (PAM), pulse position modulation (PPM), and pulse width modulation (PWM) are three common methods of digitization of a voice (analog) signal.

2.2 PULSE AMPLITUDE MODULATION

Sampling is the first step in digitization of an analog signal. In this step, an analog signal will be divided vertically into many impulse signals. In pulse amplitude modulation, the amplitude of the sampled (impulse) signal is proportional to the amplitude of the analog signal at the moment of sampling. These impulse signals are separated from each other in an equal time interval (sampling period).

The simplest method of sampling an analog signal is to send the signal through an electronic ON/OFF switch in order to have a signal during the ON time and no signal during the OFF time. The OFF time actually separates two consecutive impulse signals. However, the simplest method is not always the best and most efficient method. Conversion of an analog to a digital signal is not the only issue in transmission of an analog signal over a digital medium. The reconstruction of the transmitted analog signal at the receiver with the same characteristics is also a major issue. Reconstruction of the original signal is directly dependent on how many impulse signals the ON/OFF switch has produced. If there are not enough impulse signals, reconstruction of the original signal with the same characteristics on the receiver side is almost impossible.

In this regard, the question is how many impulse signals are enough to reconstruct the original signal? The sampling theorem is the answer to this question. The sampling theorem, also known as the *Nyquist* or *Nyquist–Shannon theorem*, refers to determination of the sampling frequency that would enable us to reconstruct the transmitted analog signal with the same characteristics on the receiver side. The sampling theorem states that to recover all the information that is contained in a signal during the sampling process, the signal must be sampled at least two times more than the highest frequency of the signal. In other words, the sampling frequency $f_s \geq 2 f_{max}$, is also known as a Nyquist frequency.

Example 1: Find the sampling frequency of a communication system that transmits signals with a frequency range of 2.4 to 3.6 kHz.

Solution: According to the sampling frequency theorem, the sampling frequency should be equal to or greater than twice the maximum frequency of the source signal. Therefore,

$$f_s \geq 2 f_{max}$$
$$f_s \geq 2 \ (3.6 \ \text{kHz})$$
$$f_s \geq 7.2 \ \text{kHz}$$

Electronic Switching in the Sampling Process

A regular ON/OFF switch is unable to sample an analog signal, such as voice, with hundreds of thousands of samples per second needed to transmit over a digital communication system. A transistor is a better device than a regular ON/OFF switch for sampling. Figure 2–1 shows an electronic circuit for sampling an analog signal.

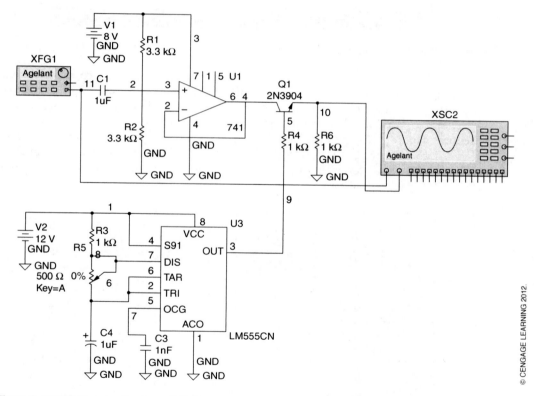

Figure 2–1: Sampling circuit using a 555 timer as an electronic switching device

This sampling circuit will sample a 400-Hz, 4-$V_{p\text{-}p}$ sine wave, as shown in Figure 2–2.

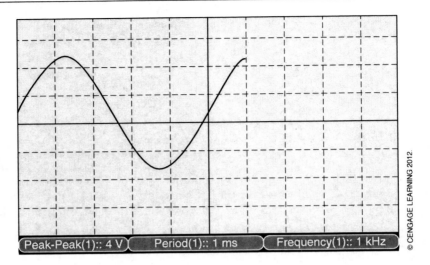

Figure 2–2: A 400-Hz, 4-V$_{pp}$ input sine wave signal

The buffer (unity gain negative feedback circuit) is used to stabilize the input signal and remove the negative half of the input sine wave. The output of the buffer is connected to a bipolar NPN transistor while its base is connected to a variable oscillator circuit to be turned ON and OFF according to the sampling theorem. Figure 2–3 shows the sampled signal with (a) sampling frequency less than ($f_s = 743.5$ Hz), (b) close to ($f_s = 860.9$ Hz), (c) 1.5 times ($f_s = 1.2$ kHz) greater than, and (d) about seven times ($f_s = 5.38$ kHz) greater than the Nyquist frequency ($f_s \geq 2\ (400\text{Hz}) = 800$ Hz).

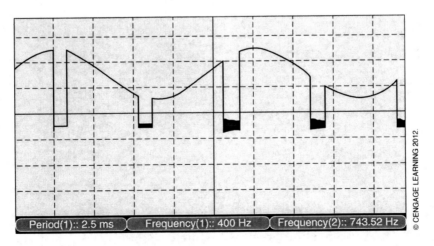

Figure 2–3a: Sampling frequency is less than the Nyquist frequency

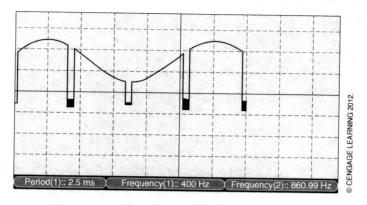

Figure 2–3b: Sampling frequency is close to the Nyquist frequency

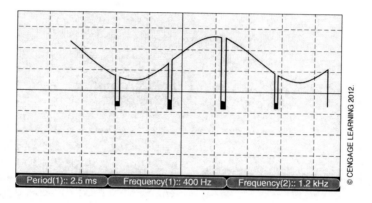

Figure 2–3c: Sampling frequency is 1.5 times greater than the Nyquist frequency

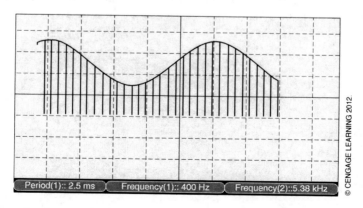

Figure 2–3d: Sampling frequency is about seven times greater than the Nyquist frequency

Aliasing

Aliasing is a phenomenon in which a signal is sampled at a rate less than the sampling frequency rate. Aliasing is actually the distortion of the original signal, which means the reconstructed signal is different from the original signal. This is because the high frequency components of the signal have been mistakenly taken for the lower frequency components.

Over- and Under-sampling

Over- and under-sampling refers to sampling a signal at a rate higher and lower than the Nyquist sampling frequency rate, respectively. No additional information will be saved by oversampling, but it will help us to design a cheaper anti-aliasing filter or higher resolution Analog-to-Digital (A/D) to Digital-to-Analog (D/A) conversion. Undersampling is used to sample a bandpass filter and for this reason is also called *bandpass sampling*.

Holding

Holding in the sampling process will help us convert an analog signal to a digital signal in order to transmit it over a digital medium. The digital signal is actually a series of impulse signals with different amplitudes. These amplitudes must be recognized in order for the analog signal to be correctly reconstructed in its original shape at the receiver. To recognize these amplitudes, we need to hold the amplitude of each impulse signal constant during the sampling period after generating a specific impulse. Employing a capacitor at the output of a sampling circuit will help us hold each sample for a sampling period of time. There are many encoding methods in data communication. These are explained in detail in Chapter 4. Return to Zero (RZ) and Non Return to Zero (NRZ) are two basic encoding methods. The sampled signal and sample-and-hold signal, respectively, are representations of these two encoding methods.

Figure 2–4a shows the circuit that can be used to sample and hold an analog signal and Figure 2–4b shows a sample-and-hold signal. The two buffers at the first and last stages of this circuit are used to stabilize the input signal before it enters into the sample/hold circuit and before transmission.

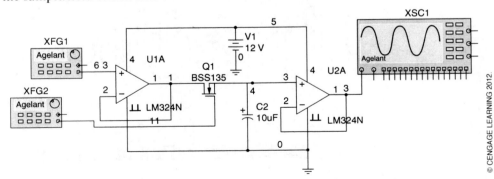

Figure 2–4a: Sample-and-hold circuit

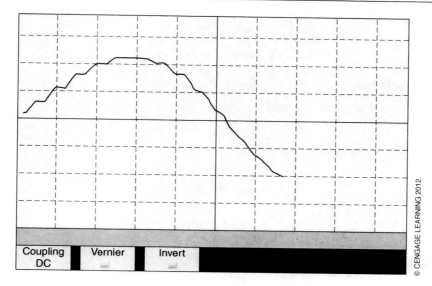

Figure 2–4b: Sample-and-hold signal

Dynamic Range

Dynamic range (DR) is the ratio of the largest to smallest (nonzero) magnitude of a variable quantity. In PCM, the DR is the ratio of the maximum voltage value to the minimum voltage value in the sampling process. The minimum voltage value also refers to the quantum value or resolution. PCM converts an analog signal to a digital signal and therefore, it is an analog-to-digital converter (ADC). The dynamic range for an ADC is equal to:

$$\text{DR (dB) ADC} = 20 \log \left(\frac{2^n}{1} \right) = 20 \, n \log 2 = 6.02 \, n \tag{2.1}$$

Equation (2.1) shows that in PCM, the number of bits in the data depends on the dynamic range. The maximum quantization error (Q_E) and the resolution of the PCM signal also depend on the dynamic range according to the following equations:

$$Q_E = \frac{V_{max}}{2 \, DR} \tag{2.2}$$

$$\text{Resolution} = \frac{V_{max}}{DR} \tag{2.3}$$

Example 2: A ± 3-V human voice is sampled by a PCM and transmitted over a digital transmission line. Find the following values if the DR = 48 dB:

    a. The minimum sampling frequency
    b. The minimum number of bits required to be coded by a PCM

    c. The minimum number of levels

    d. Resolution

    e. Quantization error

Solution:

    a. The range of the human voice is between 300 and 4000 Hz. Therefore,

$$f_s \geq 2\, f_{max} = 2\,(4000) = 8000 \text{ Hz}$$

    b. $DR(dB) = 6.02\, n$

$$48 = 6.02\, n$$
$$n = 8 \text{ bits}$$

    c. $M = 2^n$

$$M = 2^8 = 256 \text{ levels}$$

    d. Resolution $= \dfrac{V_{max}}{DR}$

$$DR = \frac{V_{max}}{\text{resolution}}$$

$$48 \text{ dB} = 20 \log \frac{V_{max}}{\text{resolution}} = 20 \log \frac{3V}{\text{resolution}}$$

$$\log^{-1}\left(\frac{48}{20}\right) = \frac{3V}{\text{resolution}}$$

$$251.18 = \frac{3V}{\text{resolution}}$$

$$\text{Resolution} = \frac{3V}{251.18} = 0.0119$$

    e. $Q_E = \dfrac{\text{resolution}}{2} = \dfrac{0.0119}{2} = 0.0059$

2.3 PULSE POSITION MODULATION

In pulse position modulation, or PPM, the position of the sampled pulse is shifted according to the amplitude rate change of the modulating signal while the amplitude of the sampled pulse remains at a constant level. PPM is mostly used in noncoherent modulation in optical communication where the coherent phase modulation and detection is expensive and difficult. Another application of PPM is in low power, narrowband, and long wavelength radio frequency (RF) channels that are subject to the flat fading phenomenon. Figure 2–5 shows a PPM signal.

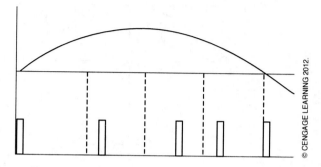

Figure 2.5: Pulse position modulation (PPM) signal

2.4 PULSE WIDTH MODULATION

In pulse width modulation, the pulse width or pulse duration of the modulated signal varies by the rate of change in the analog signal, while its amplitude remains the same. The modulated signal has the greatest pulse width at the peak value of the analog signal and will decrease as the amplitude of the analog signal decreases. There are varieties of electronic circuits to generate a PWM signal. PWM is a popular method of modulation that is used widely in the electronic industry. Figures 2–6a and 2–6b show a PWM circuit and PWM signal, respectively.

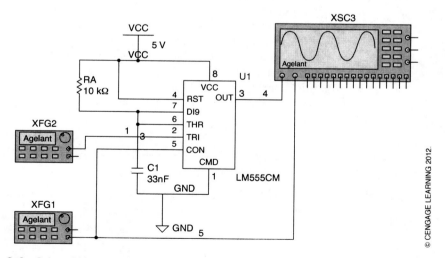

Figure 2–6a: Pulse width modulation (PWM) circuit

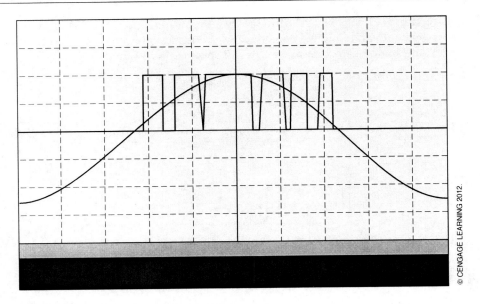

Figure 2–6b: PWM signal

2.5 QUANTIZATION

Simply put, quantization is the conversion of an analog signal into a series of impulse signals (digital signal). Sampling and quantization are also called an *A/D conversion process*. The quantization process is actually a sample/hold process with a preassigned number of holdings (equal levels) in order to recover the original signal and to be separated from channel noise. For example, if a sampled signal was held eight times during the quantization period, then we must assign 3-bit data for each quantized level, which also represents the amplitude of this sampled signal. Every sample signal will fit between two different quantized levels. The amplitude of a quantized level is called the *step size*. Step size is calculated by dividing the net value of the analog signal ($V_{max} - V_{min}$) by the number of quantized levels or steps N.

$$\text{Step size } (\Delta S) = \frac{(V_{max} - V_{min})}{N} \tag{2.4}$$

Example 3: What would be the step size of a 2-V peak value of an analog signal that is quantized in eight levels? Show the quantized signal graphically.

$$\Delta S = 2 \text{ V}/8 = 0.25 \text{ V}$$

The graphical representation of the quantized signal is shown in Figure 2–7.

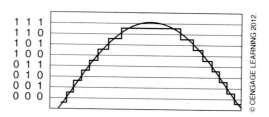

```
1  1  1
1  1  0
1  0  1
1  0  0
0  1  1
0  1  0
0  0  1
0  0  0
```

© CENGAGE LEARNING 2012.

Figure 2–7: An eight-level quantized signal

Table 2–1 shows all eight quantized levels along with their amplitude values and the assigned 3-bit digital data values.

Quantized level (S)	Amplitude value (V)	Assigned 3-bit data
1	0.25	000
2	0.5	001
3	0.75	010
4	1.00	011
5	1.25	100
6	1.50	101
7	1.75	110
8	2.00	111

Table 2–1: The voltage assignment for an eight-level sample-and-hold signal

© CENGAGE LEARNING 2012.

To recover the original analog signal without error depends on where each sampled signal is located. In general, any of the sampled signals is either on one of the steps or is between two steps. In the first case, the signal will be recovered without error. In the second case, there would be an error in the recovery process because there is no amplitude or binary data assigned to that specific sampled signal. An amplitude value halfway between the quantization levels is usually assigned in a D/A converter. In Example 3, the halfway amplitude would be 0.125 V. By this arrangement, any sampled signal that is lying between two quantized levels (s), is actually lying between $(-\Delta s/2)$ and $(\Delta s/2)$. This error is known as quantization error and will be calculated as discussed in Section 2.6.

2.6   SIGNAL-TO-QUANTIZATION NOISE RATIO

The signal-to-quantization noise ratio (SQNR) is the ratio of the input signal power to the quantized noise signal power.

Signal Power

If the input analog signal is a sinusoidal signal, then the average value of the signal power is equal to:

$$[\text{Signal power}]^2 = 1/\pi \int_{-\pi/2}^{\pi/2} \sin^2 e \; de \qquad (2.5)$$

$$= 1/\pi \int_{-\pi/2}^{\pi/2} \frac{1}{2}(1 - \cos 2e)\, de$$

$$= 1/\pi \left[\frac{1}{2}\left(e - \frac{1}{2}\sin 2e \right) \right]_{-\frac{\pi}{2}}^{\frac{\pi}{2}}$$

$$= \frac{1}{2}$$

Quantized Noise (Error) Power

The quantization error is the difference between the amplitude of the original signal (the signal that needs to be quantized) and the amplitude of the quantized signal. The average value of the quantized power is actually the same as the probability density of the quantized signal between $[-\Delta s/2, \Delta s/2]$.

$$[\text{Quantized error power}]^2 = 1/\Delta s \int_{-\Delta s/2}^{\Delta s/2} e^2 de \qquad (2.6)$$

$$= 1/\Delta s \left[\frac{1}{3} e^3 \right]_{-\Delta s/2}^{\Delta s/2}$$

$$= \frac{\Delta S^2}{12}$$

Since the converting of the step size (ΔS) into the quantization interval (0.25 V) is equal to $2^{-(n-1)}$ where n is the number of bits representation of a quantized level, then:

$$[\text{Quantized error power}]^2 = [2^{-2(n-1)}]/12 \qquad (2.7)$$

Therefore, the SQNR would be:

$$SQNR = 10 \log [\text{signal power/quantized noise power}]$$
$$= [(\tfrac{1}{2})/(2^{-2(n-1)})/12] = 10 \log [6/(2^2 \, 2^{-2n})]$$
$$= 10 [\log (6/4) - \log (2^{-2n})]$$
$$SQNR \text{ (dB)} = 10 [0.17609 + 2n (0.3)] = 1.76 + 6n \qquad (2.8)$$

Example 4: How many bits are required to represent levels of an analog signal that is quantized in eight levels by a pulse code modulation circuit? Also find the signal-to-quantized error ratio.

Solution:

$$n \text{ (Number of bits)} = \log_2 \text{ (number of levels)}$$
$$n = \log_2 (8) = \log_2 (2^3) = 3 \log_2 (2) = 3 \text{ bits}$$
$$\text{SQNR} = 1.76 + 6 \, (n) = 1.76 + 6 \, (3) = 19.76 \text{ dB}$$

2.7   NONUNIFORM QUANTIZATION

The amplitude variation is not uniform for all input signals. As a result, the above uniform quantization method may or may not maintain an SQNR of 30 dB or better as required by the International Telegraph and Telephone Consultative Committee (CCITT) standards. The nonuniform amplitude variation causes different step sizes (ΔS) in the quantization process. Therefore, there is a smaller step size at the bottom of the signal and a larger step size at the top of the signal. To resolve this problem, the step size of the larger amplitude variation must be compressed and the lower amplitude variation must be expanded in order to maintain uniform step size. The compression and expansion of the input analog signal is called *companding*. In uniform companding, the PAM signal will be encoded and digitally compressed at the transmitter side before transmission and will be expanded and then decoded at the receiver side. Depending on the type of companding, the compression and expansion circuits may or may not be included in the encoder and decoder circuits.

2.8   DIGITAL SIGNAL LEVELS

As discussed earlier, an analog signal is transformable to a digital signal through the sampling and quantization process. The human voice is an analog signal with frequencies between 300 and 3000 Hz. A frequency of 3000 Hz is also the maximum frequency required for a telephone conversion. According to the sampling theorem, the sampling rate for the human voice should be at least two times the highest frequency or $f_s \geq 2 \, f_{max} \geq 2 \, (3000) \geq 6000$ Hz. To receive a clear voice signal, experiments have shown that the human voice needs to be sampled at a frequency of 8000 Hz. If the human voice is sampled at 8000 Hz and quantized by eight bits, then the minimum bit rate or the capacity of a one-voice-equivalent channel would be:

$$(8 \text{ bits}) \, (8000 \text{ per sec}) = 64,000 \text{ bits-per-second (bps)} = 64 \text{ Kbps}$$

The 64 Kbps is the base level of the digital signal, which is called *digital signal level zero* or *DS-0*. The DS-0 has capacity of one voice channel or one telephone line. The duration (time period) of the DS-0 signal is $1/8000 = 125$ microseconds. The DS-0 rate may support one 64 Kbps, five 9.67 Kbps, ten 4.8 Kbps, or twenty 2.4 Kbps clear channels.

The higher digital signals, DS-1, DS-2, DS-3, and DS-4, can be achieved by multiplexing lower digital signals, as shown in Figure 2–8.

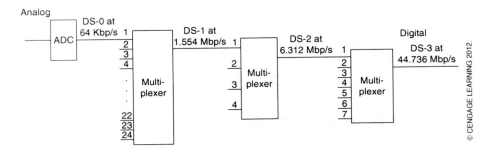

Figure 2–8: Generating digital signal (DS) levels

According to Figure 2–8, the capacities of higher levels of digital signal are as follows:

- The capacity of a DS-1 signal is equal to the capacity of 24 DS-0 channels plus framing or synchronization bits that are required to allow the receiver to recognize which bit belongs to which of the 24 channels. Therefore, the capacity of DS-1 = [(24 channels) (8-bit word each) + 1 bit for synchronization] [(sampling frequency of 8000)] = [(193 bits) (8000 per second)] = 1544000 bps = 1.544 Mbps. Duration of the DS-1 signal remains the same as that of the DS-0 signal, at 125 microseconds. The DS-1 channel is also called the *T-1 line*.
- The capacity of the DS-2 signal is 6.312 Mbps, which is equal to the capacity of four DS-1 channels (6176 Mbps) plus control and framing bits. Duration of the DS-2 signal remains the same as that of the DS-0 signal, at 125 microseconds. The DS-2 channel is also called a *T-2 line*.
- The capacity of the DS-3 signal is 44,736 Mbps, which is equal to the capacity of seven DS-2 channels (44,184 Mbps) plus control and framing bits. The DS-3 channel is also called a *T-3 line*.
- The capacity of the DS-4 signal is 274,176 Mbps, which is equal to the capacity of six DS-1 channels (268,416 Mbps) plus control and framing bits. The DS-4 channel is also called a *T-4 line*.

Example 5: Calculate the overhead value of a *T-1 line*.

Solution:

$$\text{Overhead} = \frac{1 \text{ bit of synchronization}}{192 \text{ bits of data}} \times 100 = 0.52\%$$

2.9   DELTA MODULATION

The size of the bandwidth in a communication system is one of the parameters that directly affect the efficiency of the system. A larger bandwidth means lower efficiency. Conversion of an analog signal by pulse code modulation requires a large bandwidth because PCM will quantize the entire input signal.

One way of improving the efficiency of a communication system is by lowering the size of its bandwidth. This can be done by comparing the amplitude of the present sample with the amplitude of its past sample. That means the quantization of an analog signal will take place based on the change in the signal from sample to sample, instead of the absolute value of the signal at each sample. In this method the bit stream of the output signal is a series of bit 0 or bit 1. If the amplitude of the present sample is smaller than the amplitude of its past sample, a bit 0 will appear at the bit stream of the output signal. If it is greater than the amplitude of the past sample, a bit 1 will appear. This method is called *delta modulation* (DM) and it is the simplest type of differential pulse code modulation (DPCM).

While oversampling should be avoided in all three types of PCM, in DM oversampling is required in order to accurately predict the next input condition. This means the sampling frequency must be much higher than twice that of the highest input frequency. In DM, the transmitted data will be reduced to a 1-bit data stream; therefore, the size of transmission will be reduced and the communication system will be able to transmit more data.

How a Delta Modulator Operates

The sampled analog signal (the present sample) will be compared by a comparator with the past sample and fed back to the comparator through an integrator/predictor. If both present and past samples are the same, then the output of the comparator which calculates their difference (the error term) will be zero and a bit 0 will be assigned to the output of the comparator by a 1-bit quantizer circuit. Otherwise, a bit 1 will be assigned. At the receiver side, the modulated signal will go to an integrator and then a low-pass filter (to smooth the signal). Based on the linear operation characteristic of an integrator, the two integrators of the modulator and demodulator in the DM can be combined into one. They can be placed at the front of the 1-bit quantizer to make the circuit simpler without altering the input and output signals. Sigma-delta modulation is a modulation that employs only one integrator.

Figure 2–9 shows the block diagram of a delta modulator/demodulator.

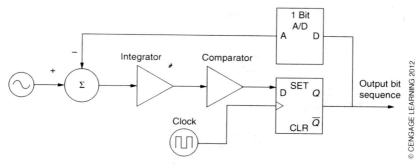

Figure 2–9: Block diagram of a delta modulator/demodulator

Slope Overload and Granular Noise

There are two major issues in performance of the DM: slope overload distortion and granular noise. Slope overload distortion refers to the rapidly rising input signals. In this case, the frequency of the input signal is faster than the speed of the modulator and therefore the step size is too small to respond to part of the input signal that has a steep slope. Slope overload distortion can be reduced by increasing the step size. Granular noise refers to the difference between the step size and the assigned sampled voltage. This means that the step size is too large when the input signal has a small slope. Granular noise can be reduced by lowering the step size. Figure 2–10 shows the effect of slope overload in a DM.

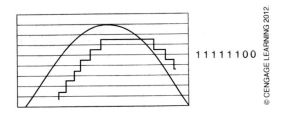

1 1 1 1 1 1 0 0

© CENGAGE LEARNING 2012.

Figure 2–10: The effect of slope overload in a delta modulator

2.10 SIGNAL-TO-QUANTIZATION NOISE RATIO IN A DELTA MODULATOR

To find the maximum output signal-to-quantization noise ratio (SQNR) for a DM, the following criteria need to be taken into account:

1. There is no slope overload for sinusoidal signals.
2. The quantization error is uniformly distributed over $[-\Delta s/2, \Delta s/2]$ where Δs is the step size of quantization (staircase).
3. The standard deviation of the error is flat during the time interval of 1-bit quantization.
4. The message bandwidth (f_c) is equal to the bandwidth of the low-pass filter in a delta modulator decoder.

Based on the above criteria, the average quantization noise power at the output of the low-pass filter is equal to:

$$P_{(QN)} = \frac{(\Delta s)^2}{12} \frac{f_c}{f_b}$$ (2.9)

The maximum signal power of a DM is:

$$P_{(N)} = \frac{(\text{amplitude of the signal})^2}{2} = \frac{\left(\dfrac{\Delta s}{2}\right)^2 f_b^3}{8\pi^2 f_c^3}$$ (2.10)

The SQNR in a delta modulator is the ratio of the maximum noise power $(P_{(N)})$ and the average quantization noise power $(P_{(QN)})$ and it is equal to:

$$SQNR = 10 \log \left[\frac{P_{(N)}}{P_{(QN)}} \right]$$

$$= 10 \log \left[\frac{3 f_b^3}{8\pi^2 f_c^3} \right]$$

$$= 10 \log [0.036 \, (f_b/f_c)^3] \qquad (2.11)$$

Example 6: Find the SQNR of a delta modulator if the bandwidth of the input signal is 2.4 kHz and the bit rate is 56 Kbps.

Solution: Use Equation (2.11) for f_c = 2.4 kHz and f_b = 56 Kbps

$$SQNR = 10 \log [0.036 \, (56/2.4)^3]$$
$$SQNR = 26.6 \text{ dB}$$

Example 7: What would the signal-to-noise ratio be if the bit rate doubles?

Solution: The new bit rate would be 112 Kbps and the SQNR would be:

$$SQNR = 10 \log [0.036 \, (112/2.4)^3]$$
$$SQNR = 35.63 \text{ dB}$$

The SQNR will be increased by 9 dB for doubling the bit rate. This means that faster communication comes with higher SQNR. Therefore, a better mechanism than DM is needed to keep the SQNR at a low value while increasing the speed of communication.

2.11 ADAPTIVE DELTA MODULATION

As discussed earlier, to reduce slope overload the step size must be increased. However, increasing the step size will result in higher granular noise. Therefore, improving one problem in DM will result in worsening another problem. Employment of a controllable step-size mechanism that is sensitive (adapts itself) to the slope of the sampled signal in DM will resolve this problem. This type of DM is called *adaptive delta modulation* (ADM).

SUMMARY

To transmit or receive an analog signal such as the human voice to or from a digital system is impossible unless it is converted into a digital signal. Sampling is a process of conversion of an analog signal to a digital signal. The analog signal at the transmitter and receiver sides must be identical to have the exact signal on both sides. According to the sampling or Nyquist theorem, an analog signal must be sampled at least at double its highest frequency in order to recover the transmitted signal in its original shape. In fact, sampling is modulating an analog signal with a digital signal.

Pulse code modulation or PCM is a modulation technique that is used for the transmission of an analog signal over a digital channel. Pulse amplitude modulation (PAM), pulse code modulation (PPM), and pulse width modulation (PWM) are three fundamental types of PCM. Each of these techniques has advantages and disadvantages over the other types and they are used in different applications.

Sampling and quantization are two processes that will convert an analog signal to a digital signal. Different digital signal levels have been developed for faster digital communication. A multiplexer is used to combine some lower level signals to develop a higher level signal. A specific noise, called *quantum noise*, will be generated in the process of the sampling and quantization. The quantum-to-signal noise ratio needs to be fully understood in designing a digital communication system. The delta modulator and adaptive delta modulator were developed based on differential PCM (DPCM) to improve the efficiency of digital communication.

Review Questions

Questions

1. In pulse amplitude modulation (PAM), the amplitude of the sampled signal is _____ to the analog input signal.

2. In pulse position modulation (PPM), the amplitude of the modulated signal remains at the _____ level, while its pulse position is proportional to the _____ of the analog input signal.

3. In pulse width modulation (PWM), the amplitude of the modulated signal remains constant, while its pulse width is proportional at the _____ of the input signal.

4. Which digitized signal forms a Return to Zero (RZ)?

5. A PWM signal can be created by EXoring (Exclusive ORing) a _____ wave and a _____ wave.

6. Pulse code modulation (PCM) consists of _____ at its transmitter side.

7. Pulse code modulation (PCM) consists of _____ at its receiver side.

8. In order to satisfy Nyquist sampling criteria, the analog signal at the input is band-limited by a _____, then sampled with _____ equal to or larger than _____ its bandwidth.

9. The quantizer and the encoder circuits together perform the fundamental function of _____.

10. What is the function of a low-pass filter in the pulse code modulator?

11. What is the function of a quantizer in the pulse code modulator?

12. What is the function of the decoder at the receiver side of the PCM?

13. What is the main difference between uniform and nonuniform quantization?

14. How can slope overload be minimized in a delta modulator?

15. What are the advantages of an adaptive delta modulator with respect to a delta modulator?

16. What is the relationship between the bandwidth of the input signal and the SQNR in a delta modulator?

Continues on next page

17. What is the main function of the encoder and decoder in pulse code modulation?

18. Quantization noise and bit error rate (BER) are two important parameters that determine the overall accuracy of the _____ analog signal.

19. In order to generate a(n) _____ of the transmitted levels by disregarding the channel noise, the magnitude of noise must be less than _____ that of the step size.

20. The major disadvantage of transmitting an analog signal over a digital channel is the _____ requirements.

Problems

1. Calculate the overhead value in a DS-2 signal.

2. Determine the signal-to-quantization noise ratio in a pulse code modulator that uses 16- and 32-bit word quantization.

3. Calculate the step size of a 12-mV peak-to-peak analog signal when it is quantized at 16 levels.

4. What is the minimum sampling frequency of an analog input signal when its frequency is 5.6 kHz?

5. A telephone line carries voice signals with a bandwidth of 300 Hz to 3300 Hz. What is the minimum sampling frequency?

6. The voice line in Question 5 is quantized by eight bits (256 levels of voltage) and the sampling frequency is 8 kHz. What is the transmission speed for this basic voice line (or DS-0 signal)?

7. Find the SQNR of a delta modulator if the bandwidth of the input signal is 3.6 kHz and the bit rate is 64 kbps.

8. Determine the bandwidth of the input signal of a delta modulator if its SQNR is 40 dB with a 128 Kbps bit rate.

9. Calculate the noise power of a quantized signal with step size S = 2.

10. What would be the output signal power of Problem 9 if the quantized signal has eight levels?

11. How many DS-0 signals are required to have a DS-1 signal? Calculate the transmission speed of a DS-1 signal.

12. How many DS-1 signals are required to have a DS-2 signal? Calculate the transmission speed of a DS-2 signal.

13. How many DS-2 signals are required to have a DS-3 signal? Calculate the transmission speed of the DS-3 signal.

14. What is the efficiency of DS-1 transmission (remember there is one frame bit for DS-1 transmission)?

15. A ± 1 V_{p-p} human voice is sampled by a PCM and transmitted over a digital transmission line. Find the following values if the DR = 24 dB:

   a) The minimum sampling frequency
   b) The minimum number of bits required to be coded by a PCM
   c) The minimum number of levels
   d) Resolution
   e) Quantization error

Pulse Amplitude Modulation

Introduction: Write a brief description of the pulse amplitude modulation process and its application.

Parts and Equipment

- 1-kΩ resistor
- 1-kΩ potentiometer
- 1.0-µF and 0.3-µF capacitors
- 555 timer
- 1RFZ46N or similar N-MOS transistor
- Power supply (12 V-DC)
- Function generator
- Oscilloscope

Construct the PAM circuit shown in Figure 2–11.

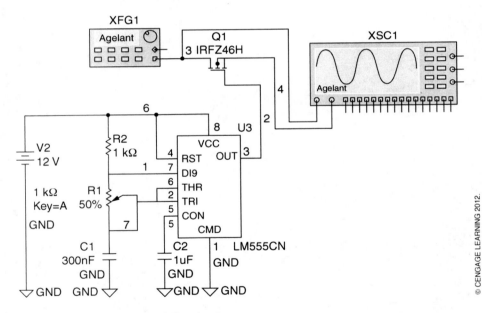

Figure 2–11: Pulse amplitude modulation circuit

1. Set the function generator at 5 V_{p-p} and 200 Hz.

2. Set the potentiometer on its 50% range.

3. Calculate the sampling frequency using the following equation:

$$f_s = \frac{1.44}{(2\ R1\ +\ R2)\ C1}$$

4. Save the sampling signal from the oscilloscope screen.

5. Change the potentiometer value from 50% to: 5%, 20%, 40%, 60%, 80%, and 90% and calculate the sampling frequency for each case and compare their sampling signals (number of samples per cycle) with the sampling signal in Step 4.

6. Summarize your findings in Table 2–2.

Resistor value of the potentiometer (%)	Sampling frequency (kHz)	Number of samples per cycle
5		
20		
40		
50		
60		
80		
90		

Table 2–2: Comparison of the resistor value of the potentiometer, sampling frequency, and number of samples per cycle in the sampling circuit

7. Set the sampling frequency to twice the input signal frequency $(2)(200Hz) = 400\ Hz$ by changing the resistor value of the potentiometer and find what the sampling signal looks like.

Questions

1. What would happen if you change the capacitance of the C_1 capacitor instead of changing the resistance of the potentiometer?

2. What is the purpose of the C_2 capacitor?

3. Is the input signal in this experiment recoverable if you set the sampling frequency on 400 Hz? Explain your answer.

Optional: Construct the sample-and-hold circuit of Figure 2-4a and compare it with the PAM circuit in this experiment and write your conclusion.

Continues on next page

Conclusion

Pulse Width Modulation

Introduction: Briefly describe pulse width modulation, its function, and applications.

Parts and Equipment

- One 3.3-kΩ, two 0-kΩ, and one 33-kΩ resistors
- 500-Ω potentiometer
- One 1.0-nF, one 2.2-nF, two 10-nF, and one 100-μF capacitors
- Two 555 timers
- Power supply (5 V-DC)
- Function generator
- Oscilloscope

Construct the pulse width modulation circuit shown in Figure 2–12.

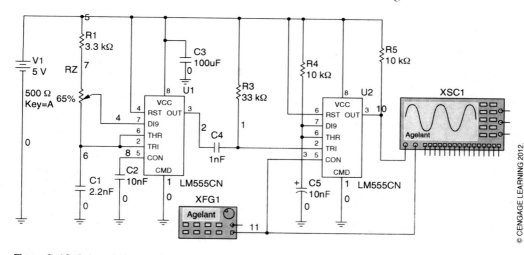

Figure 2–12: Pulse width modulation circuit

1. Set the function generator at 3 V$_{p-p}$ and 300 Hz.

2. Set the potentiometer on 30%, 40%, 50%, and 60% of its maximum value and use the following equation to calculate the frequency of each setting:

$$f_s = \frac{1.44}{(2\,R2\,+\,R1)\,C1}$$

3. Save the PWM signal of each case and compare them with the base value of 30%.

Continues on next page

Questions

1. Why do we need two 555 timers? Can you design a PWM with a single 555 timer? How? Show your circuit.

2. Which case among the four experimental cases will provide the fastest PWM signal for speedy transmission of the modulated signal? Why? Explain your answer.

3. In what field of the electronic industry is PWM used the most?

Conclusion

3

Signal Conversion, Case Two: Digital-to-Analog Signal Conversion

Objectives

After completing this chapter, students should be able to:

Describe digital-to-analog signal conversion and the related modulation techniques.

Discuss the performance analysis of a QAM system.

Describe modems and cable modems.

The majority of data storage in data communication, such as computers, is in digital systems. Therefore, data that needs to be transmitted in these types of systems are in digital format. For transmission of data in digital format over an analog transmission line such as a telephone line, we need to convert the digital data into an analog signal. This process is called modulating digital data into an analog signal. An analog signal is defined by three parameters or characteristics: amplitude, frequency, and phase. A change in any of these three parameters will result in an analog signal with different characteristics. These characteristics of an analog signal will help us modulate digital data by changing one or more of the parameters. Amplitude shift keying (ASK), frequency shift keying (FSK), and phase shift keying (PSK) are three methods for modulating digital data into an analog signal. Quadrature amplitude modulation (QAM) is a modulation technique for higher bandwidth efficiency using both the amplitude and phase of the carrier signal.

3.1 AMPLITUDE SHIFT KEYING (ASK)

Binary amplitude shift keying (BASK), where the data sequence has two levels, is the simplest type of ASK. One of the levels is usually zero and the other level has either plus or minus amplitude values. In BASK modulation, the output carrier signal will switch between two amplitude values according to the input data sequence levels, while its frequency and phase remain the same. Figure 3–1 shows a block diagram of a BASK signal through a multiplier that modulates the input data sequence in a digital signal format and the carrier signal generated by a local oscillator.

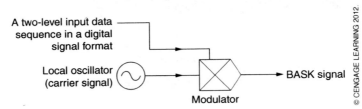

A two-level input data sequence in a digital signal format

Local oscillator (carrier signal)

Modulator

BASK signal

© CENGAGE LEARNING 2012.

Figure 3–1: Process of generating a BASK signal

A two-level ASK signal can be defined by the following equation:

$$E(t) = \begin{cases} E_0 \cos(\omega t + \varphi), & 0 \le t \le T, \text{ for binary logic "1"} \\ 0, & \text{otherwise, for binary logic "0"} \end{cases} \quad (3.1)$$

where E_0 is the amplitude of the output carrier signal and T is the bit duration. ASK also is called *ON-OFF keying* (OOK) which refers to the condition of having a signal and not having a signal. The ON and OFF states are also called *mark-space states*. Figure 3–2 shows the input data sequence and the output carrier signal.

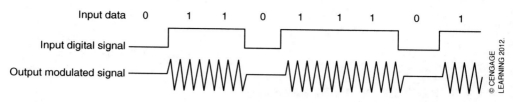

Figure 3–2: The input data sequence and the output modulated signal in the BASK modulation technique

One of the disadvantages of ASK is its wide bandwidth, which results from sharp discontinuity at the transition points. To resolve this problem, we may band-limit the ASK signal by applying a band-limiting (bandpass) filter before transmission. However, this will result in receiving a band-limited version of the original signal at the receiver end. To recover the original data sequence, we need to have a two-step demodulation process at the receiver end, as follows:

1. The recovery process of the band-limited digital signal
2. The regenerating process of the input data sequence

M-array ASK signal refers to the modulation of a multilevel data sequence. For example, the 4-ASK is a 4-level modulation technique for two bit data sequences (00, 01, 10, and 11). The lowest level of the 4-ASK signal is assigned to the data sequence of "00," and the higher levels are assigned to the data sequences of "01," "10," and "11." Therefore, each level of the 4-ASK signal contains 2-bit data.

The bandwidth of the modulation is proportional to the signal rate of communication. The proportional factor is equal to $(1 + k)$ where the k parameter depends on the modulation process and its value varies between 0 and 1 $(0 < k < 1)$. Therefore, the relation between the bandwidth and the signal rate is defined by the following equation.

$$BW = (1 + k) f_s \qquad (3.2)$$

The bit rate is also proportional to the signal rate and the proportional factor(s) represents the number of bits per signal.

$$f_b = s \cdot f_s \qquad (3.3)$$

Example 1: Find the bit rate of a data communication system with bandwidth of 5 kHz. Assume ASK is used to modulate the transmitted data and the k parameter is equal to 1.

Solution: In the ASK method of modulation only one bit is modulated per signal. Although a higher level is possible, it is not implemented in ASK. It is used in other modulation of modulation methods such as QPSK and QAM. Therefore, the bit rate in ASK is equal to the baud rate since $(s = 1)$. Use Equation (3.2) to find the signal rate and Equation (3.3) to find the bit rate of the communication system.

$$5000 \text{ Hz} = (1 + 1) f_s$$
$$f_s = 2500$$
$$f_b = s \cdot f_s = (1) (2500) = 2.5 \text{ Kbps}$$

3.2 FREQUENCY SHIFT KEYING (FSK)

In frequency shift keying, the frequency of the carrier signal will be varied to represent the input data sequence. For example, there would be two carrier frequencies of f_1 and f_2 that represent the bit "1" and bit "0" for a two-level input data sequence. In other words, for a two-level input data sequence, the transmitter employs two oscillators with oscillation frequency of f_1 and f_2, where only one of them will be connected to the output at any one time. In practice, both f_1 and f_2 are selected to be integer multiples of the bit rate, which leads to phase continuity and, therefore, bandwidth reduction. Figure 3–3 shows a block diagram of a frequency shift keying modulation technique.

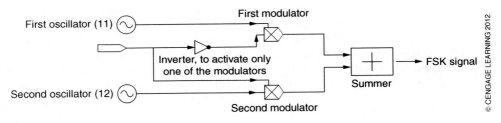

Figure 3–3: Block diagram of an FSK modulation technique

If we represent the two carrier signals by $E_1(t) = E_0 \cos (\omega_c + \omega_d)t$ and $E_2(t) = E_0 \cos (\omega_c - \omega_d)t$, for bit "1" and bit "0," respectively, then the composite FSK signal at the output of the modulator will be as follows:

$$E(t) = E_0 \cos (\omega_c \pm \omega_d)t \qquad (3.4)$$

where ω_c is the carrier frequency, ω_d is the peak frequency deviation, and $f_2 = \omega_c + \omega_d$, and $f_1 = \omega_c - \omega_d$. The input data sequence and the composite FSK signal for $f_2 > f_1$ is shown in Figure 3–4.

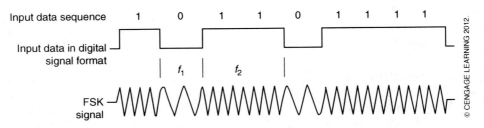

Figure 3–4: The input data sequence and the output signal in the FSK modulation technique

Since the peak frequency deviation (ω_d) is inversely proportional to the bit duration of the input binary (T), and the bit duration is inversely proportional to $(f_2 - f_1)$, then the peak frequency is directly proportional to $(f_2 - f_1)$ as follows:

$$\omega_d = \frac{(f_2 - f_1)}{2} \tag{3.5}$$

$$T = \frac{1}{2(f_2 - f_1)} \tag{3.6}$$

Then the minimum peak frequency deviation in the FSK modulation is one quarter of the bit duration.

$$\omega_d = \frac{1}{4T} \tag{3.7}$$

The bandwidth of the FSK signal depends on the frequency of the two carriers $(f_2$ and $f_1)$ and the bit duration.

$$BW = (f_2 - f_1) + 2\left(\frac{1}{T}\right) = 2\left(\omega_d + \frac{1}{T}\right) \tag{3.8}$$

Example 2: Determine the peak frequency deviation, bit duration, and bandwidth of the FSK signal if the frequencies of two local oscillators are 1 kHz and 1.2 kHz.

Solution:

$$\omega_d = \frac{(f_2 - f_1)}{2} = \frac{1.2 \text{ kHz} - 1 \text{ kHz}}{2} = 100 \text{ Hz}$$

$$T = \frac{1}{2(1.2 \text{ kHz} - 1 \text{ kHz})} = 2.5 \text{ msec}$$

$$BW = (1.2 \text{ kHz} - 1 \text{ kHz}) + 2\left(\frac{1}{2.5 \text{ msec}}\right) = 1000 \text{ Hz}$$

The FSK signal can be demodulated synchronously (coherently) and asynchronously (noncoherently). In synchronous demodulation, the FSK signal will be divided into two separate signals with frequencies of f_1 (lower channel or space) and f_2 (upper channel or mark), and two local oscillators are used to demodulate these two signals synchronously. The output signals of the demodulators will pass through a low-pass filter and then feed to a decision-making circuit. The decision-making circuit will examine the two modulated signals and select the signal that is most likely one of the input signals. The decision-making circuit will also reshape the selected signal to be in its original shape. A synchronous demodulation circuit is more complex than an asynchronous demodulation circuit. Figure 3–5 shows a block diagram of the synchronous (coherent) demodulation of an FSK signal.

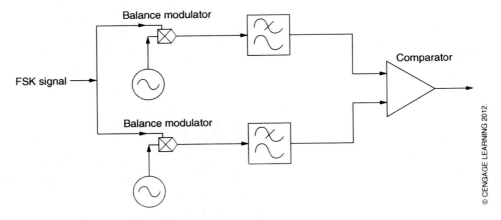

Figure 3–5: Block diagram of the FSK synchronous demodulator

In an asynchronous modulator, each separated signal will first go through a band-pass filter and then through an envelope detector circuit for an asynchronous demodulator. Finally, like the synchronous demodulator, the decision-making circuit will select one of the input signals and also will reshape it to be the same as its original shape. Figure 3–6 shows a block diagram of an asynchronous demodulator for an FSK signal.

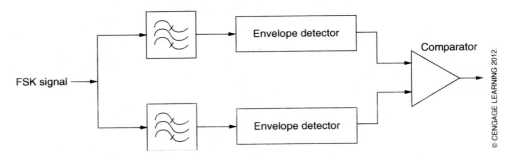

Figure 3–6: Block diagram of the FSK asynchronous demodulator

3.3 MINIMUM SHIFT KEYING (MSK)

A sharp transition between bits "1" and "0" and vice versa in the input data sequence creates a signal with a sideband that may extend from the carrier signal. This phenomenon will cause interference between two adjacent channels in any digital radio communication. Using minimum shift keying, which is a modified FSK modulation (also known as a modified PSK modulation), may resolve this problem. There is no phase discontinuity in MSK, and the frequency for bit "0" is 1.5 times the frequency of bit "1." In other words, the frequency difference between bit "1" and bit "0" is equal to half of the data rate. In fact, MSK is a special type of continuous-phase FSK (CPFSK) where two carrier frequencies are at an exact 180-degree phase difference. This characteristic of the MSK allows the FSK signals to be coherently orthogonal and provides the minimum bandwidth. The Gaussian MSK (GMSK) is actually a modified version of MSK that uses a premodulation filter to reduce the bandwidth of a baseband pulse train before modulation takes place.

3.4   PHASE SHIFT KEYING (PSK)

In the phase shift keying modulation technique one signal is the inverse of the other signal. This means the phase of the carrier signal will shift (180 degrees) or will not shift according to the value of the bits (0 or 1) in the input data, while the amplitude and frequency remain constant (constant envelope). PSK requires the same bandwidth as ASK and is more resistant to noise than ASK and FSK. The disadvantage of PSK is the requirement of a reference carrier signal at the receiver end to determine the exact phase of the transmitted signal to be able to determine whether the transmitted signal is in mark or space condition. Varieties of PSK are used in wireless and amateur radio communications. The input data sequence and the PSK signal are shown in Figure 3–7.

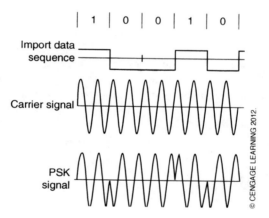

Figure 3–7: PSK signal

Binary Phase Shift Keying (BPSK)

Binary phase shift keying is the simplest type of PSK modulation. In BPSK modulation, the phase of the output modulated signal alternates between zero (in-phase, I) and 180 degrees (out-of-phase, Q) with respect to the local oscillator (carrier signal). The modulated signal is in-phase with the local oscillator if the bit in the input data sequence is "1" and it is out-of-phase if the bit is "0." In BPSK, the input data sequence will go through a bipolar comparator to be changed to positive/negative signals. Bit "0" will change to a negative signal and bit "1" to a positive signal. These two signals will trigger the modulator to change the phase of the carrier signal 0 or 180 degrees, respectively, to construct the BPSK signal. At the end of this process, a bandpass filter will smooth the BPSK signal. Figure 3–8 shows a block diagram of the BPSK modulating technique.

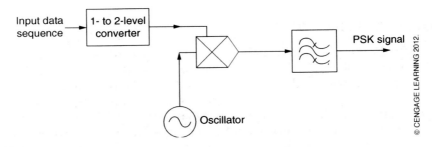

Figure 3–8: Block diagram of BPSK

The mathematical representation of the BPSK signal can be shown by the following equation:

$$E(t) = \sqrt{\frac{E}{T}} \, \cos{[\omega_c t + (1-n)\pi]} \tag{3.9}$$

where $n = 0$ and 1. E is the energy per bit and T is the bit duration.

By substituting $n = 0$ and 1, we will able to find the equations for bit "0" and bit "1" and locate them on an I-Q diagram (constellation diagram).

$$E_0(t) = \sqrt{\frac{E}{T}} \, \cos{(\omega_c t + \pi)} = -\sqrt{\frac{E}{T}} \, \cos{(\omega_c t)}, \, n = 0 \text{ or bit "0"} \tag{3.10}$$

$$E_1(t) = \sqrt{\frac{E}{T}} \, \cos{(\omega_c t)}, \qquad\qquad n = 1 \text{ or bit "1"} \tag{3.11}$$

A constellation is a visual presentation of the PSK modulation technique as a vector in a two-dimensional format with in-phase (I) as a horizontal axis, and quadrature phase (Q) as a vertical axis. The amplitude of the modulated signal is shown by the length and angular frequency of the vector. The vector can rotate counterclockwise about the horizontal axis. In the BPSK modulation technique, bit "1" represents a vector in-phase with the carrier signal and bit "0" represents a vector 180 degrees out-of-phase with the carrier signal. Equations (3.10) and (3.11) show that the signals for bit "0" and "1" are the inverse of each other on a circle with radius of $\sqrt{E/T}$. Figure 3–9 shows the BPSK constellation diagram.

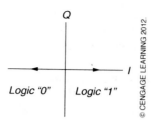

Figure 3–9: BPSK constellation diagram

The minimum channel bandwidth in BPSK is equal to the input data rate.

Quadrature-Phase Shift Keying (QPSK)

Multilevel PSK is the advanced PSK modulation technique for a higher bit rate or smaller bandwidth. Multilevel PSK usually is shown by 2^n – PSK where parameter n is the bits per signal element. For example, there are four signal levels and 2 bits per signal element if $n = 2$. PSK with four signal levels is called *quadrature-PSK* (QPSK). In the QPSK modulation technique, the input data sequence is fed into a bit splitter (a serial-to-parallel converter) to split the even-numbered and odd-numbered bits of the 2-bit data sequence. The even-numbered bits will go through a bipolar comparator before being assigned to the quadrature balanced modulator. The odd-numbered bits will go through the same procedure but will be assigned to the in-phase balanced modulator. Then, both in-phase and quadrature-phase modulated signals will be combined by a summing circuit to create a QPSK signal. A bandpass filter will smooth the QPSK signal. Figure 3–10 shows a block diagram of the QPSK modulating technique. Splitting 2-bit input data in QPSK to two 1-bit input data is nothing but converting a QPSK modulator to two BSPK modulators.

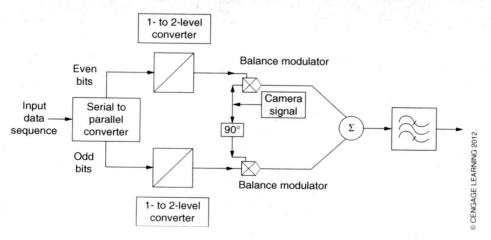

Figure 3–10: Block diagram of QPSK

The bit rate in QPSK is twice the bit rate in BPSK for the same bandwidth, or the bandwidth of the QPSK is half of the bandwidth of the BPSK for the same bit rate. That is, the minimum channel bandwidth in QPSK is equal to half of the binary data rate.

The mathematical representation of QPSK is shown in the following equation:

$$E(t) = \sqrt{\frac{2E_s}{T_s}} \cos\left[\omega_c t + (2n - 1)\frac{\pi}{4}\right] \tag{3.12}$$

where E_s and T_s are energy per symbol and symbol duration, respectively, and $n = 1, 2, 3,$ and 4 represents the four signal levels in the QPSK. To come up with the two-dimensional signal-space equations of these four-level signals from Equation (3.12) the following criteria should be taken into consideration.

1. The in-phase channel is shown by $\cos(\omega_c t)$.
2. The quadrature-phase channel is shown by $\sin(\omega_c t)$.
3. The QPSK is a balanced modulator, which means $E(t)$ is the same in both the I channel and the Q channel.
4. The 2-bit per signal representation is as follows: $1 = 00, 2 = 01, 3 = 10,$ and $4 = 11$.
5. The four phases of the 2-bit signals are as follows: $\pi/4, 3\pi/4, 5\pi/4,$ and $7\pi/4$.
6. Bits "0" and "1" are represented by negative and positive signals, respectively.

I	Q	Voltage level	Phase value	Signals in the I-channel	Signals in the Q-channel	QPSK vector in the constellation diagram
0	0	−1 and −1	$\frac{\pi}{4}$	$-\cos(\omega_c t)$	$-\sin(\omega_c t)$	3rd quadrant
0	1	−1 and +1	$\frac{3\pi}{4}$	$-\cos(\omega_c t)$	$-\sin(\omega_c t)$	2nd quadrant
1	0	+1 and −1	$\frac{5\pi}{4}$	$\cos(\omega_c t)$	$\sin(\omega_c t)$	4th quadrant
1	1	+1 and +1	$\frac{7\pi}{4}$	$\cos(\omega_c t)$	$\sin(\omega_c t)$	1st quadrant

Table 3–1: Mathematical presentation of the four signal levels in the QPSK constellation diagram

The QPSK constellation diagram shown in Figure 3–11 can be constructed using Table 3–1.

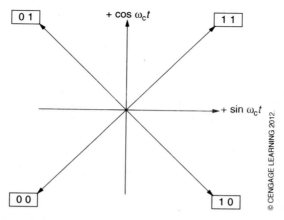

Figure 3–11: QPSK constellation diagram

Higher levels of PSK have been developed for faster bit rates or smaller bandwidths. Eight signal levels (8-PSK) and 16 signal levels (16-PSK) are samples of higher level PSK that would modulate 3 bits per signal element and 4 bits per signal element, respectively. 8-PSK uses eight phases of 0, ±45, ±90, ±135, and 180 degrees. The modulating circuitry becomes more complex as the signal levels increase.

Example 3: The bandwidth of a data communication system that uses the QPSK method of modulation is 4 MHz. Find the baud rate and bit rate. Assume the k parameter value is 1.

Solution: As was discussed earlier, there are four signals at the output of a QPSK modulator (signals with four different phases). This means QPSK modulates two bits of information per signal. Therefore, the r value is 2 and the baud and bit rate values are as follows:

$$BW = (1 + k) f_s$$
$$4 \text{ MHz} = (1 + 1) f_s$$
$$f_s = 2 \text{ Mbaud}$$
$$f_b = s \cdot f_s = (2)(2 \text{ Mbaud}) = 4 \text{ Mbps}$$

3.5 DIFFERENTIAL-PHASE SHIFT KEYING (DPSK)

Clock recovery is an important issue in synchronous (coherent) communication. Any phase difference in a signal between a receiver and transmitter will cause additional phase error, and consequently bit error, in a data communication system. One way to resolve this problem is to synchronize the receiver's clock with the transmitter clock. Differential-phase shift keying is an alternative method to resolving this problem. The difference between PSK and DPSK is in their encoding of the input data sequence. PSK encodes the input data sequence in-phase, while DPSK encodes it in the phase difference between successive bits or symbols. This means that there would be a phase change in the modulation signal if the two successive bits in the input data sequence are different (0 to 1 or 1 to 0), and no phase changes if the successive bits are the same. DSPK is called *conventional DPSK* (CDPSK) if the phase difference ($\Delta\varphi$) is in the set of $\{0, \pi\}$ and symmetrical DPSK (SDPSK) if the phase difference is in the set of $\{-\pi/2, \pi/2\}$. Experiments have shown that SDPSK has a better bit-to-error ratio than CDPSK in the presence of intersymbol interference (ISI). DPSK is widely used in long-haul optical communication.

3.6 QUADRATURE AMPLITUDE MODULATION (QAM)

Bandwidth is an important parameter in any communication. Efficient use of the bandwidth is not just necessary but it is absolutely required in a well-designed communication system. Experiments have shown that a mixture of ASK and PSK, called *quadrature amplitude modulation* (QAM), has much better bandwidth efficiency and a much lower probability error than PSK. The input data sequence is carried simultaneously by the amplitude and phase of the carrier signal in the QAM modulation technique. In other words, in QAM modulation the amplitude and phase of the carrier signal varies according to the input data sequence. High spectrum efficiency in QAM provides more channels in a given limited bandwidth. Varieties of QAM have been developed with higher transmission rates. For example, 64-QAM is capable of transmitting 27 Mbps, which is equivalent to 6 to 10 analog channels or one HDTV signal over a 6-MHz bandwidth. The 4-, 8-, 16-, 32-, 64-, and 256-QAM are among the varieties of QAM with different symbol rates. Symbol rate is defined as the ratio of bit rate over the number of bits per symbol. Table 3–2 shows the symbol rate for different types of QAM based on an 80-Kbps bit rate (an 8-bit sampler voice signal at 10 kHz). The minimum channel bandwidth in QPSK is equal to one-third of the binary data rate.

Type of QAM	Number of bits per symbol	Symbol rate (K symbol/sec)
2-QAM (BPSK)	1	80
4-QAM (QPSK)	2	40
8-QAM	3	26.6
16-QAM	4	20
32-QAM	5	16
64-QAM	6	13.3
256-QAM	8	10

Table 3–2: Symbol rate for different types of QAM

With these advantages of QAM, compared to other modulation techniques that we reviewed earlier, come two main disadvantages. QAM is more suscepti-ble to noise and it requires having a linear amplifier on the transmitter side. The linear amplifier consumes a lot of power, which makes it unattractive in mobile communication.

The input data sequence in QAM will be divided into three bit sequences and fed into three channels of in-phase (I), quadrature (Q), and control (C) chan-nels. The C-channel will determine the magnitude of the signal in both I ($I = C \cos(\omega_c t)$) and Q channels ($Q = C \sin(\omega_c t)$. Bits in both the I channel and Q channel will go through separate 2- to 4-level converters, a low-pass filter, and then be fed into a balanced modulator. Finally, the modulated signal from the I and Q chan-nels will be mixed by an adder and a bandpass filter to smooth the QAM signal. Figure 3–12 shows a block diagram of the QAM modulation technique. Table 3–3 summarizes the characteristics of the three channels that help to construct its con-stellation diagram.

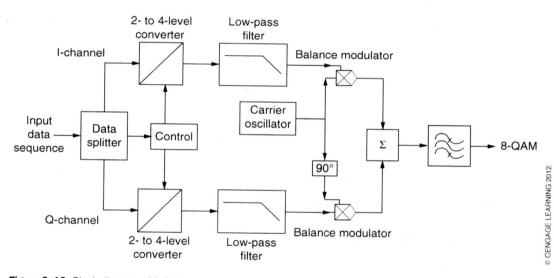

Figure 3–12: Block diagram of 8-QAM

The magnitude of the C channel is equal to 0.414 for the lower circle and 1.0 for the upper circle. The magnitude of each point (phasor) on the constellation diagram (E) is equal to $\sqrt{(mag\ of\ I)^2 + (mag\ of\ Q)^2} = \sqrt{(0.414)^2 + (0.414)^2} = 0.586\ or = \sqrt{(1)^2 + (1)^2} = 1.414\ f$ or lower and upper circles respectively and the phase of each point (φ) is equal to $tan^{-1} \dfrac{mag\ of\ Q}{mag\ of\ I} = -135\ or\ +135$ depending on the location of the point on the constellation diagram.

Input data sequence			Signals in the I channel	Signals in the Q channel	C	E	φ	Quadrant
0	0	0	$-0.414 \cos (\omega_c t)$	$-0.414 \sin (\omega_c t)$	0.414	0.586	−135	3rd
0	0	1	$-1.0 \cos (\omega_c t)$	$-1.0 \sin (\omega_c t)$	1.0	1.414	−135	3rd
0	1	0	$-0.414 \cos (\omega_c t)$	$+0.414 \sin (\omega_c t)$	0.414	0.586	+135	2nd
0	1	1	$-1.0 \cos (\omega_c t)$	$+1.0 \sin (\omega_c t)$	1.0	1.414	+135	2nd
1	0	0	$+0.414 \cos(\omega_c t)$	$-0.414 \sin (\omega_c t)$	0.414	0.586	−45	4th
1	0	1	$+1.0 \cos (\omega_c t)$	$-1.0 \sin (\omega_c t)$	1.0	1.414	−45	4th
1	1	0	$+0.414 \cos(\omega_c t)$	$+0.414 \sin (\omega_c t)$	0.414	0.586	+45	1st
1	1	1	$+1.0 \cos (\omega_c t)$	$+1.0 \sin (\omega_c t)$	1.0	1.414	+45	1st

© CENGAGE LEARNING 2012.

Table 3–3: Magnitude, phase, and location of phasors in the QAM constellation diagram

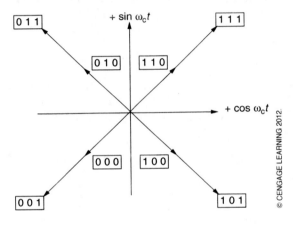

Figure 3–13: QAM constellation diagram

Performance Analysis of a QAM System

Like in any other modulation technique, bandwidth efficiency, clock recovery, carrier recovery, and symbol rate are important factors to be considered in analyzing the performance of QAM.

Bandwidth Efficiency

Bandwidth efficiency, which is also called *spectral efficiency*, is a parameter that is used to evaluate the performance of one modulation technique with respect to another. The bandwidth efficiency is the ratio of the bit rate to the minimum required bandwidth in that particular modulation technique. It is common that the bandwidth efficiency is normalized to the number of transmission bits in a 1-Hz bandwidth.

Carrier Recovery

Carrier recovery occurs when a phase-coherent reference is extracted from the carrier signal on the receiver side. For this reason, carrier recovery is also called *phase referencing*.

Symbol Rate

The symbol rate is the minimum required bandwidth to process the given data bit rate. The symbol rate is the same as the baud rate and is measured in baud or symbols per second. Therefore, the symbol rate in QAM is given by the following equation:

$$\text{Symbol rate} = \frac{\text{data rate at each channel}}{\log_2(n)} \tag{3.13}$$

where n is the number of bits in a symbol or level.

Clock Recovery

Clock recovery is the synchronization of the transmitter and receiver clock for precise timing to recognize how many consecutive "1s" and "0s" are in the data sequence in order to properly extract them from the receiver.

Example 4: Determine the number of bits per symbol, symbol rate, and bandwidth efficiency of 32-QAM if the bit rate of the modulation technique is 40 Mbps.

Solution: In general, a multilevel QAM is shown by 2^n-QAM, where n = number of bits per symbol. Therefore, there are 5 bits per symbol in 32-QAM, or 2^5-QAM.

The bit rate of each I or Q channel is half of the total bit rate, or $40/2 = 20$ Mbps.

$$\text{Symbol rate} = \frac{20 \text{ Mbps}}{5} = 4 \text{ million symbols per second}$$

$$\text{Bandwidth efficiency} = \frac{40 \text{ Mbps}}{4} = 10 \text{ bits symbol/Hz}$$

3.7 MODEM

A modulator–demodulator or modem is an electronic device that helps a digital system, such as a computer, transmit stored data (digital data) over an analog transmission line, such as a telephone line. A modem is either an external device that can be connected to a computer by an RS232 cable, or it is an internal device that can be plugged into a peripheral component interconnect (PCI). The transmission speed of a modem is measured either in baud rate (number of signal transmissions per second) for slow rates or in data rate or bit rate (number of bit transmissions per second) for faster rates.

The slowest rate of a modem is 300 baud and the fastest rate for uncompressed data is 57.6 Kbps. A voice/data modem has a switch that can change voice to data.

3.8 CABLE MODEM

A cable modem is actually a high-speed (about 1.5 Mbps) modem that connects a digital electronic device such as a computer to cable TV using a set-top box. The characteristics of the cable modem are summarized in Table 3–4.

Cable Modem	Downstream	Upstream
Frequency (MHz)	65–850	5-65
Bandwidth (MHz)	6 in the United States and 8 in Europe	2
Modulation technique	64-QAM or 256-QAM	QPSK or 16 QAM
Data rate	27–56 Mbps	320 Kbps–10 Mbps

© CENGAGE LEARNING 2012.

Table 3–4: Characteristics of a cable modem

A cable modem is much faster than an integrated services digital network (ISDN) or digital subscriber line (DSL), which have a data rate of less than 10 Mbps. Unlike DSL, the quality of a cable modem is not dependent on distance of subscriber from the central office.

SUMMARY

The stored information (data, image, or voice) in a digital device is in digital format. It must be converted to an analog format in order to be transmitted over an analog channel (transmission line) such as a telephone line. The conversion is nothing but modulation and demodulation of the digital signals. Several different modulation techniques have been developed to make the communication system more efficient by increasing the speed and/or lowering the bandwidth of the communication. In this type of modulation, the input data sequence will alter one or two of the three parameters (amplitude, frequency, and phase) of the carrier signal. These modulations are called *amplitude, frequency*, and *phase shift keying* and are denoted ASK, FSK, and PSK, respectively. There are also several improved versions of each of these three modulation techniques, such as 4-ASK, MSK, QPSK, and 8-PSK, which reduce the bandwidth and increase the speed of communication. QAM and its improved versions are the most popular modulation techniques. QAM is a combination of ASK and PSK. A constellation is a visual representation of PSK and QAM modulation. Modem, cable modem, ISDN, and DSL are electronic devices that convert a digital signal in digital equipment for transmission over an analog transmission line. In fact, modulation and demodulation circuitries are the major part of these devices.

Questions

1. Which of the three digital-to-analog conversion techniques (ASK, FSK, or PSK) is most susceptible to noise? Explain why.

2. In which method of modulation is the bit rate always the same as the baud rate? Explain why.

3. How many sine waves result when an ASK signal is decomposed?

4. Define a constellation diagram and its role in analog transmission.

5. List the differences between the coherent and noncoherent methods of detection.

6. What is the advantage of QAM over ASK and PSK?

7. What is the purpose of the low-pass filter in the FSK synchronous demodulator?

8. What is the purpose of the 1- to 2-level converter in QPSK?

9. What is the purpose of the bandpass filter or channel filter at the output of QAM?

10. Bandwidth efficiency is a parameter that is used to evaluate the _____ of one modulation technique relative to another.

11. What types of modulation techniques are used in cable modems?

12. What are the advantages of cable modems over DSL?

Problems

1. Calculate the baud rate for the following bit rate and modulation types:

   a) 2 Kbps and FSK
   b) 4 Kbps and ASK
   c) 6 Kbps and QPSK
   d) 36 Kbps and 16 QAM

2. What is the required bandwidth for the following cases if we need to send 4000 bps? Assume $k = 1$.

   a) ASK
   b) FSK with $2\Delta f = 4$ kHz
   c) QPSK
   d) 16-QAM

Continues on next page

3. The bandwidth of each channel of cable TV is 6 MHz. What is the available data rate for cable TV if a 64-QAM modem is being used?

4. Calculate the minimum baud rate, bandwidth, and bandwidth efficiency for the following modulation techniques:

   a) ASK with 40 Kbps
   b) FSK with 40 Kbps (mark and space have a frequency of 74 and 75 kHz, respectively)
   c) QPSK with 40 Kbps
   d) QAM with 40 Kbps

5. Calculate the minimum bandwidth, baud, and bandwidth efficiency for 20 Kbps using FSK and BPSK.

6. Determine mathematically the location of "10" for QPSK when the local (carrier) oscillator is sin ($\omega_c\ t$).

7. Determine mathematically the location of "010" (IQC) for 8-QAM when the local (carrier) oscillator is sin ($\omega_c\ t$) and the magnitude or the voltage level is 0.42 V and 1.00 V.

8. Determine the maximum number of bits we can send over a telephone line if the following modulator techniques are being used (the bandwidth of a telephone line is 4 kHz):

   a) ASK
   b) QPSK
   c) 8-QAM
   d) 64-QAM

9. What is the available data rate in a telephone line of cable TV if a 64-QAM modem is being used?

10. Determine the number of bits per symbol, symbol rate, and bandwidth efficiency of 8-QAM if the bit rate of the modulation technique is 60 Mbps.

11. Determine the peak frequency deviation, bit duration, and bandwidth of the FSK signal if the frequencies of two local oscillators are 700 Hz and 750 Hz.

12. Draw the signal representation of the data stream of "10110010" if ASK, FSK, and PSK are used to modulate the data.

Frequency Shift Keying

Introduction: Briefly describe the frequency shift keying modulation process and its application.

Parts and Equipment

- 2.2 kΩ, 22 kΩ, and 100 kΩ resistors
- 500 kΩ potentiometer
- 10 nF and 100 nF capacitors
- 2N3904 NPN-BJT or similar general purpose transistor
- 555 timer
- SPDT switch
- 8-Ω speaker
- Power supply (5 V-DC)
- Oscilloscope

1. Construct the FSK circuit as shown in Figure 3–14.

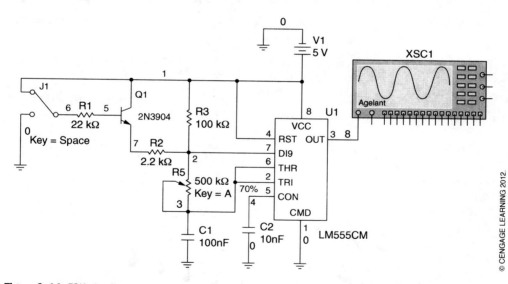

Figure 3–14: FSK circuit

2. Turn on the oscilloscope and repeatedly turn ON and OFF the single-pole, double-throw (SPDT) switch and observe how the signal on the oscilloscope behaves.

Continues on next page

3. Set the potentiometer resistance value at 20%, 40%, 60%, and 80% of its maximum value, calculate the frequency of the FSK signal, and observe the change in the signal at the oscilloscope while the SPDT switch is in its ON position.

$$f = \frac{1.44}{(2\ R5\ +\ R3)\ C1}$$

4. Replace the oscilloscope with an 8-Ω speaker and observe the sounds emitted.

Questions

1. What is the purpose of the transistor in the FSK circuit?

2. How you can turn the circuit OFF and ON electronically?

3. Find the frequency of the FSK signal when PSDT is ON and OFF. What is the importance of these two frequencies?

4. Calculate the peak frequency deviation and the bandwidth of the FSK signal using the frequencies in Step 3 and Equations (3.3) and (3.5).

5. What happens if the capacitance value of C_1 increases to double and half of its value while all resistor values remain the same?

6. What would happen if you change the capacitance of the C_1 capacitor instead of changing the resistance of the potentiometer?

Optional: Repeat this experiment with different FSK circuits and explain the differences between two circuits and their advantages and disadvantages.

Conclusion

Phase Shift Keying

Introduction: Briefly describe the phase shift keying modulation process and its application.

Parts and Equipment

- 100 Ω, 700 Ω (two), 5 kΩ, and 100 kΩ resistors
- 100 nF (two), 1 μF, and 2 μF capacitors
- 74LS04—Inverter
- 74LS86—XOR gate
- 74LS93—4-bit binary counter
- LM324N—operational amplifier
- Power supply (5 V-DC)
- Function generator
- Oscilloscope

1. Construct the PSK circuit as shown in Figure 3–15.

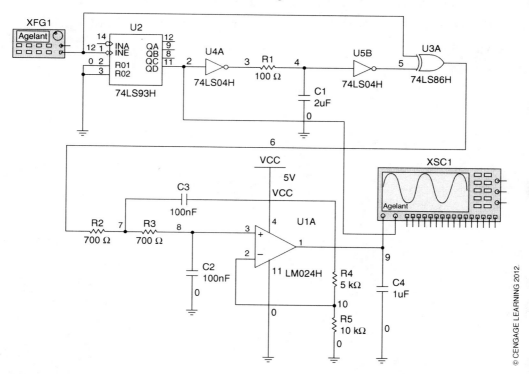

Figure 3–15: PSK circuit

© CENGAGE LEARNING 2012.

Continues on next page

2. Set the function generator at the voltage level of 20 V_{p-p} and frequency of 2 kHz.

3. Observe the PSK signal and save and include it in your report.

4. Decrease the voltage value of the function generator and explain the changes in the PSK signal.

5. Increase the voltage value of the function generator and explain the changes in the PSK signal.

Questions

1. What happens to the PSK signal if you change the resistance values of R_2 and R_3 from 700 Ω to 500 Ω? Explain your findings.

2. What happens to the PSK signal if you change the resistance values of R_2 and R_3 from 700 Ω to 800 Ω? Explain your findings.

3. What happens to the PSK signal if you change the resistance values of R_2 and R_3 from 700 Ω to 1000 Ω? Explain your findings.

4. What is the name of the op-amp circuit that is connected to the XOR?

Conclusion

Bit Splitter in PSK

Introduction: Describe bit splitting in QPSK, even and odd bit sequences, and how they get modulated.

Parts and Equipment

- 330 Ω (two) resistors
- 74LS04—Inverter
- 74LS86—XOR gate
- 74LS74—dual D-flip-flop (3)
- 74LS109—JK flip-flop
- LED (six), any color
- Power supply (5 V-DC)
- Function generator

1. Construct the bit splitter circuit as shown in Figure 3–16.

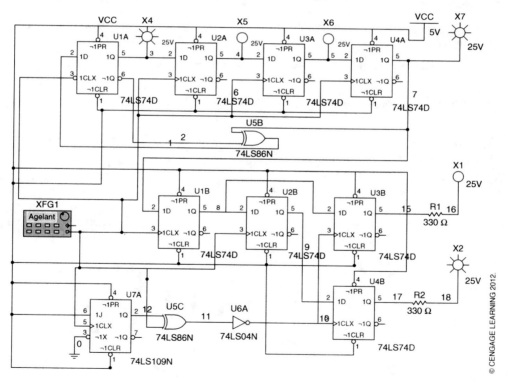

Figure 3–16: Bit splitter circuit

Continues on next page

2. Turn ON the function generator and set it on 20-Hz frequency and 9 V_{p-p}.

3. Run the circuit for a few seconds in order to see all LEDs at the top level functioning correctly. Record a random state of the top four LEDs that represent a 4-bit binary number.

4. Observe the output bits at X_1 and X_2 LEDs (these are the odd and even bits of the recorded 4-bit binary number).

5. You may adjust the frequency of the function generator in order to slow down the bit transformation from the random number generator and see the bit splitting process.

6. Connect channel one and channel two of an oscilloscope to the X_1 and X_2 LEDs and observe the signal representation of the odd and even bits.

7. Print the screen of the oscilloscope and include it in your laboratory report.

Questions

1. What is the advantage of having a random number generator rather than a simple 4-bit serial shift register?

2. How many different 4-bit numbers would be generated by the random number generator?

3. Which bit sequence (even or odd) will be modulated in-phase with a local oscillator?

4. What is the purpose of the JK flip-flop in the bit splitter circuit? What happens if it is replaced with a high active K input?

Conclusion

CHAPTER

4

Signal Conversion, Case Three: Digital Data-to-Digital Signal Conversion

Objectives

After completing this chapter, students should be able to:

Explain "data elements" and "signal elements" in data communication.

Analyze the different encoding and decoding techniques.

Discuss the differences between line and block coding methods.

Describe the functions of a scrambler and a descrambler.

Discuss transmission modes and methods.

A digital data element is a sequence of discrete numbers consisting of "0" and "1," whereas a digital signal element is a sequence of discrete pulses. Information stored electronically, such as on a computer, is a collection of a series of digital data elements. These elements do not have the characteristics of a signal and are not able to be transmitted from one point to another. Therefore, digital data cannot be transmitted in any communication system unless it is converted to a digital signal. The conversion of digital data elements to digital signal elements is called *encoding*. The process of encoding takes place at the transmitter side and the decoding process at the receiver side. The binary bit is the base value for a data element that can be either "0" or "1." One binary bit would enable us to generate only two characters. However, there are many characters in data communication, such as alphabetic characters, numbers, symbols, punctuation, and control characters. The first data element to represent a character was developed in 1874 by Jean-Maurice-Emile Baudot and is called *Baudot Code*. Baudot Code is the base for the 5-bit International Telegraph Alphabet 1 and 2 (ITA1 and ITA2), which was developed by Donald Marry for transmission of characters through a telegraph line in 1901. By advancing technology and introducing digital electronics in the communication system, much higher bits of data elements have been developed such as American Standard Code for Information Interchange (ASCII), Extended ASCII, Extended Binary Decimal Coded Interchange Code (EBCDIC), and Universal Code (Unicode).

4.1 CODING SYSTEMS

ASCII Code

ASCII code is the American version of the International Alphabet (IA-5) that was developed in 1963. ASCII code is a 7-bit coding system that presents $2^7 = 128$ characters with 96 printable and 32 unprintable characters. Printable characters include upper- and lowercase letters, numbers, punctuation symbols such as the semicolon or question mark, and other symbols such as the dollar sign and percentage. Nonprintable characters include peripheral control functions such as Enter and Escape commands. All ASCII codes come with one extra bit that is called the "parity bit." The parity bit is the most significant bit of all characters in the ASCII code. ASCII code is the perfect design code for 101 keyboards. See Appendix A for a complete ASCII table.

Example 1: Convert the word "Data" in to binary and hexadecimal representation in the ASCII code.

Solution: Table 4–1 shows the ASCII code for letters "D," "a," and "t" in both binary and hexadecimal format (see Appendix A).

Letter	Binary representation	Hexadecimal representation
D	1000100	44
a	1100001	61
t	1110100	74

Table 4.1: ASCII code for letters D, a, and t

© CENGAGE LEARNING 2012.

Therefore, the word "Data" in ASCII code is represented by "1000100 1100001 1110100 1000100" in binary or "44 61 74 44" in hexadecimal.

The Extended ASCII code is an 8-bit ($2^8 = 256$) coding system. The first 128 codes are similar to the ASCII code and the rest are assigned for special characters, including mathematical and international alphabets. The Extended ASCII code is a code for the ISO 8859-1 or ISO Latin-1 standard.

Extended Binary Decimal Coded Interchange Code (EBDCIC)

EBDCIC is an 8-bit coding system that was developed by IBM in 1964 for its mainframe and midrange computer operating system. An EBDCIC is divided into two 4-bit (nibbles) categories. The first nibble represents general characters and is called the *Zone* and the second nibble represents specific characters and is called *Digit*.

Universal Code

Increasing usage and access to computers around the world using different languages, along with the use of computers to present mathematical and scientific characters resulted in the development of a new coding system that accumulates all existing characters. The new coding system was developed in 1991 and is called universal code or simply Unicode. Unicode is a 16-bit code that represents $2^{16} = 65,356$ characters. All Unicode begins with the letter "U" and four hexadecimal digits. An ASCII code can be easily converted to a Unicode by inserting two zeroes in the hexadecimal to the left of the ASCII code. For example the letter "D" is shown by "U 0044" in Unicode. The alphabet characters of each universal language have their own block of code in the Unicode system. For instance, the block range of U0370 to U03FF is assigned to the Greek alphabet. Unicode Transformation Format (UTF) and Unicode Character Set (UCS) are the two methods of mapping in Unicode.

4.2 DATA VERSUS SIGNAL

A data element is an entity that needs to be transmitted, but does not possess any transmission characteristics. To transmit a data element, we need to convert it to a signal element. One or more signal elements may carry one or more data elements. This means the ratio(s) of the number of data elements that are carried by the number of signal elements is not the same for all cases. For example, if one signal element carries one data element, then $s = 1$. If one signal element carries two data elements, then $s = 2$, and if three signal elements carry two data elements, then $s = 2/3$. Consequently, we may say that the signal rate (f_s) is proportional to the data rate (f_b) and the r value. For $s = 1$, the signal rate is the same as the data rate, and increases if $s > 1$ and decreases if $s < 1$. Therefore, the normalized (if we set the proportional equal to 1) relationship between the signal rate and data rate is as follows:

$$f_s = f_b \cdot \frac{1}{s} \tag{4.1}$$

The signal rate, symbol rate, and baud rate are interchangeable since they all have the exact same relationship with the data rate and they are measured by baud per second.

The minimum bandwidth is proportional to the signal rate.

Example 2: Three data elements are carried by two signal elements. Show the signal-to-data relationship graphically and find the signal rate and the minimum bandwidth if the data rate is 56 Kbps. Assume the signal has two levels, +V and −V.

Solution: Figure 4–1 shows the graphical representation of a 3-data element and a 2-signal element.

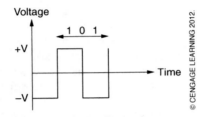

Figure 4–1: A 2-signal element that carries a 3-data element

Use Equation (4.1) to find the signal rate and the minimum bandwidth.

$$f_s = f_b \cdot \frac{1}{s} = (56 \text{ Kbps}) \, \frac{64 \text{ Kbps}}{1} = 37.33 \text{ or } 37 \text{ kbaud}$$

$$BW_{min} = f_s = 37 \text{ kHz}$$

4.3   ENCODING METHODS

The process of converting data elements to signal elements is called *encoding*. The encoding of data elements by an encoder is the last process that takes place before data is sent by a transmitter. At the receiver side, the transmitted data is decoded by a decoder to reproduce the original data. Generally speaking, an encoder/decoder is a modem in the modulation and demodulation of digital data by digital signals. Digital signals that carry some information (text, audio, or video) can be transmitted across a serial channel (link) or a parallel channel. Serial and parallel transmissions will be discussed in Section 4.6 in this chapter.

The two types of encoding are line coding and block coding, and each of them is achieved using different methods.

Line Coding

Line coding is a method of encoding that assigns a set of rules to map the change in the voltage levels that are used to represent bit "1" and "0" in a data input sequence. Line encoding is divided into five major categories: unipolar, polar, bipolar, phase, and multilevel. Each category may also be divided in to subcategories. The characteristics and applications of each category or subcategory are described below.

Unipolar Encoding

The unipolar encoding method is the simplest method, whereby either a positive or negative voltage is assigned to bit "1" and zero voltage is assigned to bit "0". The power consumption value in this method is double that of the polar method; therefore, it is a costly method to use and for this reason it does not have many applications. Figure 4–2 shows the unipolar encoding of an 8-bit data input sequence of "10110001."

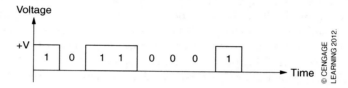

Figure 4–2: Unipolar encoded signals

Polar Encoding

The polar encoding method divides into two different encoding methods: Non Return to Zero (NRZ) and Return to Zero (RZ). The NRZ type is divided into two subcategories: NRZ-Level (NRZ-L) and NRZ-Invert (NRZ-I). The NRZ encoding method is mostly used in pulse code modulation. One of the disadvantages of the NRZ method is clock synchronization, which presents a problem when the transmitter and receiver clocks are not synchronized to each other. Having a DC component is another issue in the NRZ encoding method. The DC component will be explained in Section 4.5. A sudden change in polarity of the data link and long streams of bit "1" or bit "0" in the input data sequence are what affect the performance of the NRZ encoding method.

Non Return to Zero-Level In this encoding method, positive and negative voltages are assigned to either bit "1" or bit "0" of the data input sequence This means the voltage level changes as the input data changes from bit "1" to bit "0" or from bit "0" to bit "1". The sudden change in polarity of the data link and long streams of bit "1" or bit "0" in the input data sequence are sources of data corruption in the NRZ-L signal. Figure 4–3 shows an NRZ-L encoded signal for the input data sequence of "10110001."

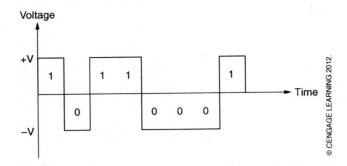

Figure 4–3: NRZ-L encoded signal

Non Return to Zero-Invert In the NRZ-I encoding method, the voltage level will change from positive to negative or negative to positive for bit "1," and there is no voltage level change for bit "0". NRZ-I encoding encounters a problem for long streams of bit "0" only and sudden polarity changes have no effect on NRZ-I. NRZ-I is also called NRZ-M (the letter M stands for mark since the voltage level changes only for bit "1" or mark). NRZ-S is the complement of NRZ-M (the letter S stands for space). NRZ-I is primarily used in magnetic tape recording. Figure 4–4 shows an NRZ-I encoded signal for the input data sequence of "10110001."

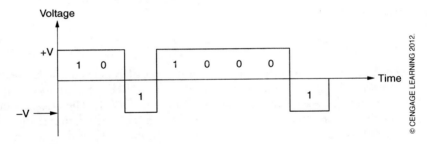

Figure 4–4: NRZ-I encoded signal

Return to Zero This encoding method uses three voltage levels: positive, zero, and negative. The voltage level will change at the beginning of any bit and returns to zero level at the midway point of the bit duration (actually, it does not have to be at the middle of the bit duration, but it is customary to show it at the middle of the bit duration). In the RZ encoded signal, the voltage level goes from positive to zero at the midway point of the bit duration for bit "1," and goes from negative to zero for bit "0". The return to zero level at the midway point of the bit duration can be used as a clock at both the transmitter and receiver sides, and may resolve the synchronization problem that occurs in the NRZ method. Resolving the synchronization comes with the use of greater bandwidth since RZ uses two signal elements to send one data element. Using greater bandwidth is the major disadvantage of the RZ method and for this reason it is not in use in digital communication. The RZ method experiences the same problem as NRZ on sudden polarity change, but has no DC component problem. Figure 4–5 shows an RZ encoded signal for the input data sequence of "10110001."

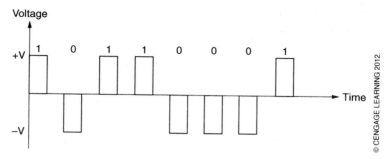

Figure 4–5: RZ encoded signal

Example 3: An input data sequence is encoded by the NRZ-I and RZ methods. Find the required bandwidth for each case if the data rate is 64 Kbps.

Solution: The ratio of the number of the data elements to the signal element for NRZ-I and RZ are: $s = 1$ and $s = \frac{1}{2}$, respectively. Use Equation (4.1) to find the bandwidth. Remember that the bandwidth is equal to the signal rate.

$$BW = s = \frac{f_b}{s} = \frac{64 \text{ Kbps}}{1} = 64 \text{ kHz} \quad \text{for} \quad \text{NRZ-I}$$

$$BW = s = \frac{f_b}{s} = \frac{64 \text{ Kbps}}{1/2} = 128 \text{ kHz} \quad \text{for} \quad \text{RZ}$$

According to the above results, the bandwidth requirement for the RZ encoding method is double the bandwidth requirement for NRZ-I, which makes the RZ method an inefficient method of encoding.

Bipolar Encoding

In this method of encoding, two alternate signal levels (positive and negative) are assigned to one bit, and the zero level signals to another bit. Therefore, there are three signal levels in the bipolar method: positive, zero, and negative. The two most common types of bipolar encoding are: alternate mark inversion or AMI, and pseudoternary, which is the complement of the AMI.

Alternate Mark Inversion In this method of encoding, an alternating signal (from positive to negative and from negative to positive) with equal amplitude represents bit "1" (mark) and no signal represents bit "0". AMI is also called *RZ-AMI* because all signals either return to zero level or are at zero level. The advantage of AMI over RZ is in its smaller bandwidth. Unlike RZ, the bandwidth requirement of the AMI is the same as NRZ. AMI does not have the DC component problem because of its alternating signals, but like NRZ has synchronization problems.

Pseudoternary Pseudoternary encoding is the complement of the AMI encoding method where the alternating signal represents bit "0" and no signal represents bit "1". A long stream of bit "0" may affect the accuracy of the AMI encoded signal.

Figure 4–6 shows (a) AMI and (b) pseudoternary encoded signals for the input data sequence of "10110001."

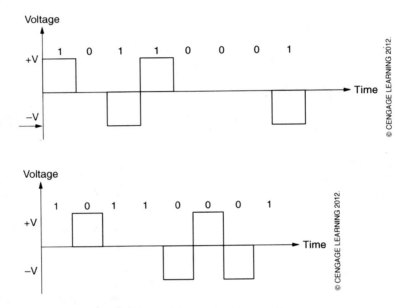

Figure 4–6: (a) AMI encoded signal; (b) pseudoternary encoded signal

Phase Encoding

In the phase or biphase encoding method, signal transition takes place at the midway point of the bit interval. Manchester phase encoding (MPE), differential Manchester encoding (DME), and delay encoding are three common types of phase encoding methods. Phase encoding methods have applications in Ethernet, satellite telemetry link, radio frequency identification (RFID), optical communication, and magnetic recording systems. The middle transition of the phase encoding method is used for the synchronization of the transmitter and receiver clocks. Unlike the NRZ method, the baseline wandering and DC component have no effect on the phase encoding method, but like the RZ method, the signal rate is two times faster than the data rate.

Manchester Phase Encoding In Manchester phase encoding or simply Manchester encoding, the transition from a high level to a low level will take place at the middle point of bit "1," and the opposite transition takes place at bit "0". It is the same as locating a pulse of one-half bit duration size at the first half and second half of the bit interval for bit "1" and bit "0", respectively. Alternatively, you may draw a Manchester encoded signal by locating upward and downward impulse size arrows at the middle of bit "1" and bit "0", respectively and then drawing a signal by following the direction of the arrows. If two adjacent arrows are in the same direction, draw one extra transition level at the beginning of the bit interval. The Manchester encoding method is also called the *biphase level encoding method*. Ethernet uses the Manchester encoding method. Figure 4–7 shows a Manchester encoded signal for the input data sequence of "10110001."

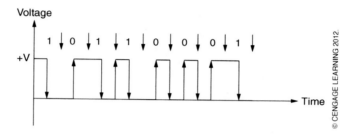

Figure 4–7: Manchester encoded signal

Differential Manchester Encoding In this method, signal transition or voltage level changes also take place at the middle of the bit interval for synchronization purposes. The difference is that the first half of the signal for bit "1" is the same as the second half of the previous bit's signal, and for bit "0" it is opposite to the previous bit's signal. In simple words, there is only one transition at the middle of the signal for bit "1" and two transitions (one at the beginning and one at the middle) for bit "0". The differential Manchester encoding method is used in Token Ring networking. Figure 4–8 shows a differential Manchester encoded signal for the input data sequence of "10110001."

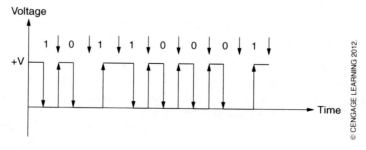

Figure 4–8: Differential Manchester encoded signal

Miller Encoding The Miller encoding method, also known as *delay encoding,* is a biphase encoding method that uses only half of the bandwidth of other biphase methods but has all of their advantages. In this method, transition occurs at the middle of the bit interval for bit "1" and at the beginning of the bit interval if there are two successive bits 0. This means, one transition at the middle for bit "1" and no transition for bit "0" unless it is followed by another bit "0". Miller encoding is used in radio signals and some smart cards used for ticketing. Figure 4–9 shows a Miller encoded signal for the input data sequence of "10110001."

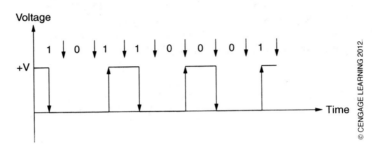

Figure 4–9: Miller encoded signal

Multilevel Encoding

Optimizing a data communication system means designing a faster delivery system along with minimum error. A faster delivery system means transmitting more data per second with smaller bandwidth. As we saw in Section 4.2, the bandwidth requirement in an encoding process is inversely proportional to r, the ratio of the data elements and the signal elements. Therefore, to increase the speed of data delivery, we must be able to assign more data elements to one unit of signal element or, in other words, reduce the bandwidth. Multilevel encoding is a method that will help us accomplish this task.

In multilevel encoding, first we partition the input data sequence into m-bit data groups, and then we break the signal element to L levels where $L = 2^m$. For example, if we partition the input data sequence to a 2-bit group, then we will be able to use a 4-level signal element ($L = 2^2 = 4$) and encode 2 bits of data at a time. In this case, the symbol rate is two times faster than the data rate.

2-Binary, 1-Quaternary 2B1Q means 2-binary (2-bit) and 1-quaternary (4-level) and is a basic type of multilevel method that uses a 2-bit group of data and is converted to one of four signal levels (symbol). This means the ratio of data elements to signal elements is 2:1 and consequently, the symbol rate is half of the bit rate. This phenomenon is known as *baud rate reduction.* The advantage of baud rate reduction is the lowering of the bandwidth on the transmission line, which leads to the reduction of the line attenuation and improvement of the immunity to near end crosstalk and noise.

2B1Q has its own unique rule: the first bit in the data element is called the *sign bit* and dictates the sign of the output quaternary level. The quaternary level will be positive if the sign is bit "1" and will be negative if the sign bit is 0. The second bit of the data element is called the *amplitude bit*, which determines the output quaternary level. The level of quaternary will be 1 if the amplitude is bit "1", and 3 if the amplitude bit is 0. Table 4–2 shows the 2B1Q encoding rules. Figure 4–10 shows a 2B1Q encoded signal for the input data sequence of "10110001."

Data element	Output quaternary
00	−3
01	−1
10	+3
11	+1

© CENGAGE LEARNING 2012.

Table 4–2: 2B1Q encoding rules

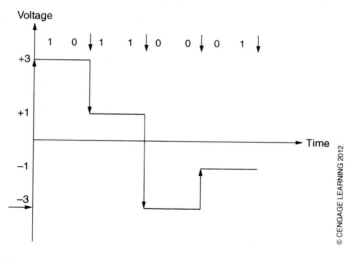

© CENGAGE LEARNING 2012.

Figure 4–10: 2B1Q encoded signal

8B/6T In this method of encoding, the input data sequence will split into an 8-bit data element and will encode with six ternary signals. The ternary signal (symbol) means 3-level signals of "+ − 0". The 8B term means there are $2^8 = 256$ data elements of 8-bits. The 6T term indicates that there are six 3-level signals and $3^6 = 729$ possible combinations to assign to the data elements, but only 256 of them will be used. Each of the 8-bit data elements has its own unique 6 ternary (6T) code. The rule for voltage level assignment for each 8-bit data element is as follows: there must be at least two voltage level transitions for synchronization purposes and an average DC voltage level of zero to eliminate polarization in the transmission line.

If the average DC voltage level of the whole transmitted data pattern is not zero, the transmitter will invert the 6T signal code(s) to create a zero average DC voltage level.

As an example, the 6T code for the data element of "00010010" or the hexadecimal number of "12," is "+ 0 + − 0 −." The complete table of 6T code is given in Appendix B. The ratio of the symbol element to the data element is 6/8 or 3/4, which means the carrier needs to run at 3/4 the rate of the input data. 8B/6T is used in 100Base-T4 standard 3-pair wire where each pair transmits two symbols.

Example 4: Show an 8B/6T encoded signal for the input data hexadecimal sequence of "1A 1F 24" or an input data binary sequence of "00011010 00011111 00100100".

Solution: From Appendix B the encoded signal for the given input data are: "0 + − + + −," "0 − + 0 + −," and "0 0 + 0 − +," and the weight for these encoded signals are: "+," "0," and "+," respectively. Since the average DC voltage level of the input data is not zero but positive, the transmitter will convert the encoded signal for input "24" to change the average DC voltage level from positive to zero. Figure 4−11 shows an 8B/6T encoded signal for the input data sequence of "00011010 00011111 00100100."

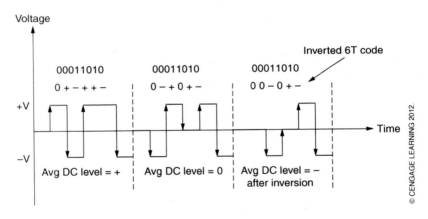

Figure 4−11: 8B/6T encoded signal

4-Dimensional, 5-Level Pulse Amplitude Modulation (4D-PAM-5) Transmission and reception of a high volume of information in Gigabit Ethernet was not possible unless an advanced method of encoding was developed. Pulse amplitude modulation-5 (PAM-5) is one of these encoding methods that is used in 100Base-T Gigabit Ethernet. PAM-5, also known as 4-dimensional 5-level PAM, employs four wires (CAT-5 where each wire operates at 125 Mbps in each direction with a total of 250 Mbps per wire and a total of 1000 Mbps or 1 Gbps in the cable) at the same time for data transmission and reception, and one wire for 4-dimensional 8-state Trellis Forward Error Detection and

Correction (FCC). In the PAM-5 encoding method each transmitted symbol represents one of the following five signal levels; -2, -1, 0, $+1$, and $+2$ volts. The ± 2 V and ± 1 V actually map to ± 1 V and ± 0.5 V, respectively. Table 4–3 shows the voltage level assignments for 2-bit data in the PAM-5 encoding method.

Bit/voltage level	Voltage level	Actual voltage level
00	−2	−1
01	−1	−0.5
10	+1	+0.5
11	+2	+1

Table 4–3: PAM-5 voltage level assignment rule

The 4-D quinary symbol (with 2-bit per symbol) provides $5^4 = 625$ signal patterns to select from four wires. These 625 signal patterns are more than enough to map all 8-bit data codes and control codes. Figure 4–12 shows the PAM-5 encoded signal for a data input sequence of "101100011000."

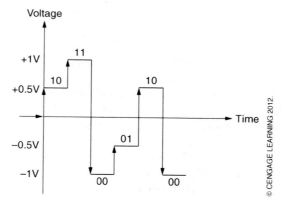

Figure 4–12: PAM-5 encoded signal

Multiline Transmission, Level-3 (MLT-3) MLT-3 is an encoding method that was designed to transmit a high data rate (100 Mbps) across an unshielded twisted pair (UTP) transmission line. MLT-3 maps one bit to one signal and uses a 3-level encoding signal of "$-V$, 0, $+V$." It takes four transitions ($0\rightarrow+V$, $0\rightarrow-V$, $+V\rightarrow0$, $-V\rightarrow0$) to complete one full cycle specifically when there is a long stream of bit "1". As a result, the maximum frequency in MLT-3 is 1/4 of those encoding methods that need two transitions to encode one bit of data. Therefore, the maximum frequency of MLT-3 is (1/4) (125 MHz) = 31.25 MHz, which is close to the FCC allowable emitting frequency. Figure 4–13 shows the state diagram for four transition levels in the MLT-3 encoding method.

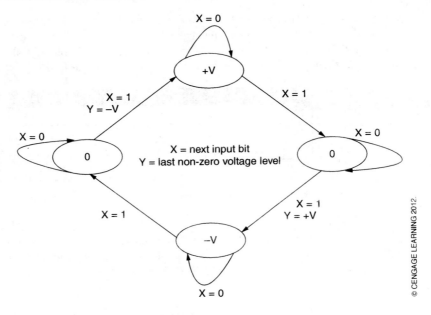

Figure 4–13: State diagram for the MLT-3 encoding signal

MLT-3 is used in FDDI copper interconnect and 100Base-TX Ethernet. MLT-3 loses its self-synchronization when there is a long stream of bit "0". A feedback shift register (FSR) that generates a pseudorandom number generator (PRNG) is added before the wave-shaping step in encoding methods such as MLT-3 that have high frequency harmonics. The PRNG circuit will reduce the regularity of the signal frequency and consequently the harmonics. A PRNG circuit is used in the following encoding experiments in this chapter: NRZ, bipolar NRZ, and RZ. The FSR is a register that shifts 1 bit at a time on each clock pulse, and an 11-bit FSR is used in 100Base-T Ethernet. Figure 4–14 shows an MLT-3 encoded signal for the input data sequence of "10110001."

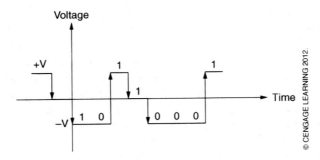

Figure 4–14: MLT-3 encoded signal

Block Coding

Block coding was developed to allow more reliable data transfer and efficient use of limited bandwidth. The 4B/5B, 5B/6B, and 8B/10B encoding methods are the most well-known block coding methods. In block coding a block of k bits will map onto a block of m bits where $m > k$. In this method, $2m - 2k$ redundant data element groups of m-bits are used for error detection and control. The unused groups are detected by the receiver as an error and will be disregarded.

4B/5B

In this method, the input data sequence will be divided into 4-bit groups and then 1 bit is inserted into each 4-bit group to become a 5-bit group. Therefore, a 4-bit data element will be inverted to a 5-bit signal element for transmission. The 4B/5B method was developed to limit the number of consecutive zeroes. Each of the 5-bit groups has at most only one leading zero starting from the left and always has at least two 1s in the bit pattern, even if the data pattern is all 0s. This means the 4B/5B method can use the NRZ-I encoding technique without losing synchronization in the case of having a long stream of bit "0" in the input data sequence. Using an extra bit for encoding a 4-bit data to a 5-bit signal will come with 25% overhead. The insertion bit will generate 16 [$(2^5 = 32) - (2^4 = 16) = 16$] 5-bit symbols that are used for controlling purposes. The control symbols are distinguishable from symbols that represent data. The advantages of the 4B/5B encoding method are in the improvement of the synchronization bit and error detection. The 4B/5B is used in 100Base-FX networks to achieve a 100 Mbps data transmission rate. Table 4–4 shows the conversion of 4-bit groups to 5-bit groups of data.

Hexadecimal number (Hex. No.)	4-bit data	5-bit signal	Hex. No.	4-bit data	5-bit signal
0	0000	11110	8	1000	10010
1	0001	01001	9	1001	10011
2	0010	10100	A	1010	10110
3	0011	10101	B	1011	10111
4	0100	01010	C	1100	11010
5	0101	01011	D	1101	11011
6	0110	01110	E	1110	11100
7	0111	01111	F	1111	11101

© CENGAGE LEARNING 2012.

Table 4–4: Conversion of a 4-bit data element to a 5-bit signal element

5B/6B

The 5B/6B encoding method is similar to the 4B/5B method but consists of mapping a 5-bit data element to a 6-bit signal element (symbol) with 32 nondata symbols for control. Twenty out of the 32 nondata symbols are used for DC-balanced transmission and 12 of them are used for transmitting data without polarization. The 5B/6B encoding method has added error-checking and is capable of detecting the invalid data patterns. Invalid patterns are those data patterns that have more than three 1s or more than three 0s in a row. The 5B/6B encoding method is used in 100VG-AnyLAN that transmits using a 25 Mbps pair of wires with a total of 120 Mbps per cable.

8B/10B

The 8B/10B method is a combination of the two block coding methods described above. The upper 3 bits of data will be encoded into a 3B/4B group and the remaining 5 bits will be encoded into a 5B/6B group. As a result, an 8-bit data element maps into a 10-bit symbol element for transmission. To maintain clock synchronization, there can be no more than six 1s or six 0s in the 10-bit groups. To make sure this criterion is met, the 10-bit groups will include either five 1s and five 0s, four 1s and six 0s, or six 1s and four 0s. Encoding of an 8-bit into a 10-bit symbol will create 256 nondata elements of 8 bits. Twelve of the 256 nondata elements that have no more than six 1s or six 0s are used as control characters to specify the start and end of the frame, link idle, skip, and other special control characters. In addition to special characters, the other advantages of the 8B/10B method are easy achievement of the bit synchronization and improvement in error detection. Two 7-bit symbols of "+ comma = 0011111" and "− comma = 1100000" are used for bit synchronization. A calculation method that is called running disparity is used to keep the number of transmitted bit "1"s and bit "0"s at the same value in order to maintain a DC-balanced transmission. The 8B/10B encoding method is used in Gigabit Ethernet, 10 Gigabit Ethernet, Fiber Channel, and ATM.

4.4   SCRAMBLER AND DESCRAMBLER

Long streams of bit "0" or bit "1" that create a constant voltage are harmful phenomena in synchronizing a communication system because there are not enough transition states. All encoding methods that we have reviewed so far face this problem. Those encoding methods that have been developed to resolve this problem have either limitations on the number of 0s or 1s in the input data sequence or are not suitable for long-distance communication.

To overcome this problem, the scrambler was developed. The scrambler is used in data communication to minimize the effect of long streams of bit "0" and bit 1, specifically bit "0" on synchronization. A scrambler, which is also called a *randomizer*, is an electronic device that inverts the data input sequence to a useful data sequence without eliminating an undesirable sequence. The scrambler sequence is the Exclusive ORing of the input data sequence and pseudorandom binary sequence (PRBS) as shown in Figure 4–15.

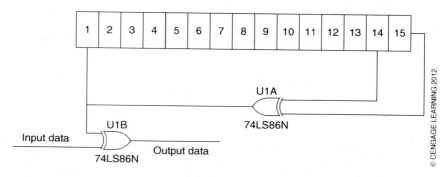

Figure 4–15: Additive scrambler using pseudorandom binary sequence

A scrambler neither encrypts the input sequence nor eliminates the probability of having a long sequence of 0 or 1. It will, however, minimize the probability of the occurrence of a long stream of successive bits. Using a scrambler in a data communication system will provide good synchronization capability, no DC component, good error detection without reduction in data rate, and finally will ease the clock recovery performance and help the transmitted data to gain the maximum power spectral density that is needed for long-distance communication. At the receiver side, the descrambler reverses this process and will extract the original data sequence from the scrambled sequence.

Bipolar with 6-zero substitution (B6ZS) and bipolar with 8-zero substitution (B8ZS) are the scrambler encoding methods that are commonly used in North America for T1 and T2 lines; whereas in Europe, the high-density bipolar 3-zero (HDB3) is in use.

B6ZS and B8ZS

These encoding methods are very similar to each other. Both of them are modified versions of the bipolar AMI encoding method. There will always be two code violations in these encoding methods. In the B6ZS method, a sixth successive zero will be replaced by a pattern of "0VB0VB." The "V" stands for violation and "B" for bipolar. If the polarity of the last mark (bit "1") before the beginning of four consecutive bit "0"s is positive, then the 6BZS code will be "0+−0−+" and if the polarity is negative, then the 6BZS code will be "0−+0+−." Figure 4–16 shows a B6ZS encoded signal for the data input sequence of "1000000101."

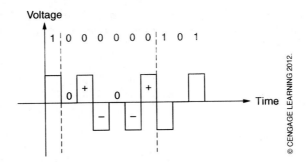

Figure 4–16: B6ZS encoded signal

Similarly, an eighth successive bit "0" will be replaced by "000VB0VB" in the B8ZS encoding method. If the polarity of the last mark (bit "1") before the beginning of the stream of bit "0"s is positive, then the 8BZS code will be "000+ −0− +" and if the polarity is negative, then the B8ZS code will be "000− +0+ −." Figure 4–17 shows a B8ZS encoded signal for a data input sequence of "100000000101." B8ZS is an encoding method for T1 lines that uses 64 Kbps for each line. As was discussed in Chapter 3, a T1 line has 24 channels of digital signal level-zero. The E1 line, which has 32 channels compared with T1, which has only 24 channels, is used in Europe and Japan. While T1 uses B6ZS and B8ZS for encoding, E1 uses the HDB3 encoding method.

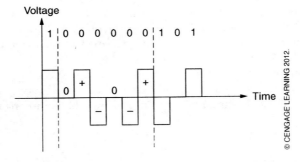

Figure 4–17: B8ZS encoded signal

High-Density Bipolar Level 3 (HDB3)

Like the B6ZS and B8ZS encoding methods, the HDB3 is a modified version of AMI but it can transmit a maximum of three successive bits of 0. If there are four bits of 0 in a row, then the HDB3 code will replace them with either "000V" or "B00V." If the DC component input sequence before the stream of four 0 bits is zero (null), the fourth successive 0 will be replaced by "000V" code, and otherwise will be replaced by "B00V." As a result, the DC component of the HDB3 will always be zero and proper clock recovery will occur. Figure 4–18 shows an HDB3 encoded signal for data input sequence of "01100001010000."

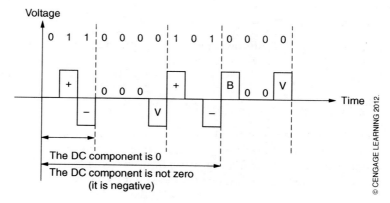

Figure 4–18: HDB3 encoded signal

4.5   RELATED TOPICS IN THE ENCODING PROCESS

The following parameters should be considered carefully before selecting any encoding method.

Self-Clocking

The transmission of data bits or symbols should come with synchronization. Some encoding methods such as Manchester or differential Manchester have embedded synchronization. Synchronization is a great help for clock signal recovery.

Error Detection

Error detection helps to recover the original data without error at the receiver side. Not all encoding methods have an error detection mechanism. Block encoding methods have reliable error detection mechanisms.

DC Component

The DC component or DC coefficient is the mean value of a periodic waveform. The existence of a long stream of bit "1" in the input data sequence will generate a constant (DC) voltage that will continuously charge the capacitor that is used as an AC coupler in a communication system. This phenomenon will result in the occurrence of bit error and imbalanced voltage between components of a communication system. To avoid such a problem, the DC component needs to be removed. In other words, the elimination of DC components, which is known as a *DC balanced waveform,* will enable the system to be AC coupled. Reliable and good encoder methods are those that have a zero DC component.

Bandwidth Reduction

As mentioned earlier in this chapter, the efficiency of an encoding technique lies in its bandwidth reduction for a given data rate. Bandwidth compression for a constant data rate is proportional to fitting more data bits in one unit of signal element.

Noise Immunity

Some channels are noisier than others and some data are more sensitive to noise than others. These are the important factors when selecting an encoding method. Elimination or at least minimization of noise must be considered carefully in selection of an encoding method for the given data and channel.

4.6   TRANSMISSION METHODS

After encoding, the final step in a data/digital communication system is transmission of the encoded signal over a transmission line. Digital signal transmission is classified into two categories or methods: serial and parallel transmissions. Since data transmission within a computer is parallel and transmission from one computer to another is mostly serial, interchanges of these two transmission methods are required for data/information transformation from one computer to another.

Serial Transmission

In this method, data will be transmitted sequentially or one bit at a time. Therefore, only one transmission line (channel) is required for the entire transformation of data. The advantage of serial communication is in the cost of wiring and its disadvantage is in a slower speed of transmission compared to the parallel mode. It is important to have a large transition density in serial transmission. Greater accuracy comes with a higher transition density. It is also very important to have a minimum number of transitions for every data group that will be transmitted serially. There are two different approaches to serial transmission: asynchronous and synchronous. Figure 4–19 shows serial transmission between two computers.

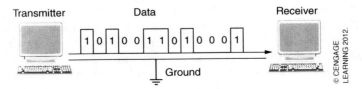

Figure 4–19: Serial transmission

Asynchronous Serial Transmission

In this approach, the transmitter and receiver are not synchronized; therefore, it is essential that both transmitter and receiver have the same communication rate. To set the transmitter and receiver communication rate to be equal, a few extra bits are added to the 8-bit (one byte) data group. The first 7 bits of the 8-bit data group represent an ASCII character and the remaining bit is the parity bit. Some extra bits are added in the data frame to indicate the start and stop of the data group. The start bit, which is usually selected to be opposite to the line-idle state, is a bit "0". The idle and start bits together will cause an idle/nonidle shift mode at the moment of data arrival at the receiver. The start bit will set the receiver clock to be synchronized with the transmitter clock and will notify the receiver of the arrival of the data group.

The stop bit, which is a bit "1," will signify the end of the data group and prepare the receiver for the reception of the next data group. The stop bit can be more than one bit. Therefore, the minimum size of the transmitted information in asynchronous transmission is 10 bits with 7 bits for data and 3 bits for the parity, start, and stop bits. In addition to the start and stop bits, asynchronous transmission comes with some gaps of different duration between groups of 10 or more bits of data to separate data frames. Transmission of extra bits along with data bits makes asynchronous transmission less efficient than synchronous transmission. The overhead for asynchronous transmission is $(2/8) \times (100) = 25\%$. Asynchronous transmission is used in keyboards, modems, and serial printers. Asynchronous transmission is used for low data rate transmission. Figure 4–20 shows the format of data transmission in the serial asynchronous transmission.

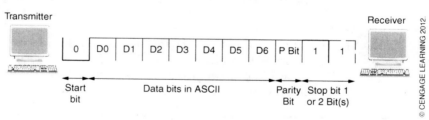

Figure 4–20: Asynchronous serial transmissions

Synchronous Serial Transmission

In this approach, data bits are transmitted continuously without start or stop bits. As a result, synchronous transmission is faster than the asynchronous method. But unlike the asynchronous method, an external clock pulse that sets the transmission rate is needed for synchronization. This additional channel will take away part of the bandwidth for nondata information. For an efficient use of bandwidth in synchronous transmission, it is customary to use an encoding method that embeds the clock signal within the input data stream, such as Manchester encoding. It is the receiver's responsibility to count the data bits and create a data group of 8 bits.

In synchronous transmission, the frame header has a predetermined bit pattern to give a unique identification for that specific frame. The bit-stuffing procedure, which is insertion of a series of bit zeroes in the frame at the transmitter side, is used when the data bit pattern in the frame and header are exactly the same. These additional zeroes will be removed from the frame at the receiver side.

Synchronous transmission is used in high speed applications and in the transmission of a large volume of information between computers. Figure 4–21 shows the transmission of an 8-bit data group in serial synchronous transmission.

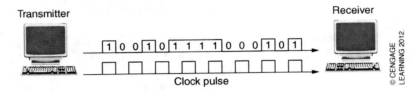

Figure 4–21: Synchronous serial transmissions

Figure 4–22 shows a typical frame format of a synchronous transmission. The frame format starts with the preamble field and ends with the frame check sequence field. The function of each field is as follows:

1. The preamble field consists of one or more bytes for synchronization purposes.
2. The start-of-frame delimiter (SFD) field signals the start of the frame.
3. The destination field contains the destination address to which the frame has to be sent.
4. The source field contains the source address from which the frame is sent.
5. The length field contains the number of bytes in the data field.
6. The data field contains actual data or information.
7. The frame check sequence (FCS) field is used for error detection.

Preamble	SFD	Destination	Source	Length	Data	FCS

Figure 4– 22: Frame format of a synchronous transmission © CENGAGE LEARNING 2012.

Parallel Transmission

In this method, each bit in the data group has its own channel but all bits are transmitted simultaneously. There exists an additional channel (the clock or strobe channel) for synchronization purposes that enables the receiver to recognize a right time to receive the transmitted data. Parallel transmission is much faster than serial transmission but comes with the high cost of wiring (eight wires versus one wire for an 8-bit data group) and more space requirements for installation. Communication ports (LPT1 and LPT2) in a computer and printer are parallel ports. Figure 4–23 shows parallel transmission between two computers.

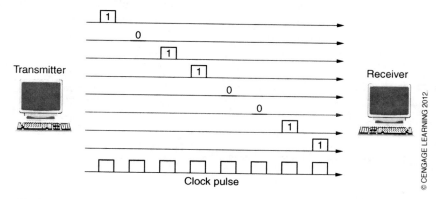

Figure 4–23: Parallel transmission

4.7 TRANSMISSION MODES

Two electronic communication devices are connected to each other by a link in order to transmit and receive signals or information packets. Transmission over a link can be classified in one of the following three modes: simplex, half-duplex, and full-duplex.

Simplex

Simplex is a one-directional transmission mode. One device can only send and the other device will only receive data. For example, a radio station sends its own radio frequency signal through the air and will not wait for any response from listeners. This mode does not have much application in data communication. Figure 4–24 shows a simplex transmission mode.

Figure 4–24: Simplex transmission mode

Half-Duplex

Half-duplex is a bidirectional transmission mode but at any given time it is only able to either send or receive data. In this transmission mode, the transmitter will send data and the receiver will respond only when the transmitter has completed data transmission. Both devices are capable of transmitting or receiving data but at any specific time only one can be functioning as a transmitter and the other as a receiver. The half-duplex mode uses the RS485 physical standard for communication.

The half-duplex transmission mode is used in the 2-wire Modbus device. A Modbus device is a serial communication protocol, which is an industrial communication standard used specifically in PLCs (programmable logic controllers). Citizens' Band (CB) radio is another example of using half-duplex transmission mode. Figure 4–25 shows a half-duplex transmission mode.

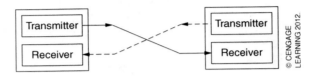

Figure 4–25: Half-duplex transmission mode

Full-Duplex

Full-duplex is also a bidirectional data transmission mode but unlike half-duplex, both devices can send and receive data at the same time (simultaneously). A handshake procedure is needed only to initiate transmission. The full-duplex transmission mode is used in 4-wire Modbus devices using the RS232C physical standard. Telephone conversation is an example of full-duplex transmission mode. It would be a very polite and respectful conversation if people used half-duplex instead of the commonly used full-duplex in their conversations.

Figure 4–26 shows a full-duplex transmission mode.

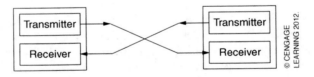

Figure 4–26: Full-duplex transmission mode

SUMMARY

The stored data in digital storage do not have any electrical property for their transmission from one point to another. Therefore, they need to be encoded—to be changed to an electrical signal—at the transmitter side before transmission, and decoded—changed back to their original shape—at the receiver side. Many different encoding methods have been developed, from the very basic, such as return to zero—which is the same as a PAM signal—to the very advanced, such as MLT-3, B8ZS, or HDB3 for faster transmission with error control mechanisms. Each of these methods has advantages and disadvantages. One method could be the best choice for one application and not a very good choice for other applications. Some encoding methods, such as Manchester and differential Manchester, have embedded synchronization features.

One of the important criteria in designing an efficient encoding system is the bandwidth reduction for given data rate. Bandwidth reduction depends on how many data bits are carried by a signal element. Self-clocking, error detection, DC component, and noise immunity are other important factors in designing or selecting an encoding method. Using a multilevel signal for the encoding process is another way of designing an efficient encoding system, but a multilevel encoding system is complex and expensive.

Encoding methods are divided into two major categories: line coding and block coding. In addition to information symbols, the block coding method produces some redundant symbols for error detection purposes and, as a result, has a better error detection capability than line coding.

Transmission of signals will take place after the encoding process. Transmission can be either serial or parallel. Serial transmission is simple and less expensive than parallel transmission, but it is slower than parallel transmission. Through advances in technology, faster serial transmission devices have been developed. Asynchronous and synchronous transmissions are two modes of serial transmission. Asynchronous transmission comes with a group of bits that includes data bits and start and stop bits. Synchronous transmission is a bit-by-bit transmission with no start or stop bits and a grouping bit to recover the transmitted characters is used by the receiver. Asynchronous transmission comes with at least 25% overhead.

Signal or data transition between two communication devices will take place in three different modes: simplex, half-duplex, and full-duplex. Simplex is a unidirectional mode, whereas half-duplex and full-duplex are bidirectional transmission modes. In full-duplex transmission both devices can transmit and receive data simultaneously, while in half-duplex transmission only one direction at a time takes place. Half-duplex and full-duplex are used in different types of Modbus devices.

Review Questions

Questions

1. What is the difference between data and signal?

2. What is the relationship between data rate and signal rate?

3. What is the difference between bit rate and baud rate?

4. Unicode is a 16-bit code that represents _____ characters. All Unicode begins with the letter _____ and four _____ digits.

5. The NRZ-L encoding method has problems with _____ change in _____ of the data link and long streams of bit "1" or bit "0" in the input data sequence.

6. What are the advantages and disadvantages of the NRZ encoding method over the RZ encoding method?

7. The AMI encoding does not have the problem of _____ because of its alternating signals.

8. How can you reduce the bandwidth for a fixed data rate?

9. Which of the line encoding methods have a smaller bandwidth?

10. Which line encoding methods have no DC component?

11. What are the differences between Manchester and differential Manchester encoding methods?

12. The Manchester encoding method is also called _____ encoding method. _____ uses the Manchester encoding method.

13. What is the advantage of a multilevel encoding method over a single-level encoding method?

14. What is the relationship between pseudoternary and AMI encoding methods?

15. What are the two major coding methods? What are the differences?

16. Explain the two types of transmission modes.

17. Explain the two types of transmission methods.

18. How many different types of communication modes exist? Describe each of them.

19. What is a scrambler? Name the encoding methods in the scrambler.

20. In the 8B/10B encoding method, 12 out of _____ nondata elements that have no more than six 1s or six 0s are used as _____ control characters to specify the _____ and _____ of the frame, _____, skip, and other special control characters.

21. How many code violations will occur in the B6ZS? How many in the B8ZS?

22. The DC component of the HDB3 will always be _____ and proper _____ will be taking place.

23. Synchronization is helpful for _____ signal recovery.

24. What is the minimum number of bits in the asynchronous transmission frame?

25. Synchronous transmission is used in the _____ applications and transmission of a _____ volume of information between computers.

26. What is the purpose of bit stuffing in synchronous serial transmission?

27. What version of Modbus is used in the half-duplex transmission mode?

28. Which physical standard is used in the full-duplex transmission mode?

29. Which physical standard is used in the half-duplex transmission mode?

30. Which transmission mode is used in TV broadcasting?

Problems

1. Use the following encoding methods to encode the ASCII code for letter M:

   a) Unipolar
   b) RZ
   c) NRZ-L
   d) NRZ-I
   e) AMI
   f) Manchester
   g) Differential Manchester

2. Use the following encoding methods to encode the data sequence of "10001101":

   a) 2B1Q
   b) MLT-3
   c) 4B/5B

Continues on next page

3. Two data elements are carried by one signal element. Show the signal-to-data relationship graphically and find the signal rate and the minimum bandwidth if the data rate is 128 Kbps. Assume the signal has two levels, of +V and −V.

4. Show the 8B/6T encoded signal for the input data hexadecimal sequence of "042264H."

5. An input data sequence is encoded by the NRZ-I and RZ methods. Find the required bandwidth for each case if the data rate is 256 Kbps.

6. Show graphically the PAM-5 encoded signal for the data input sequence of "100111000110."

7. Show graphically the B6ZS encoded signal for the data input sequence of "11000000110."

8. Show graphically the B8ZS encoded signal for the data input sequence of "100000000101." Assume the last signal level was negative.

9. Show graphically the HDB3 encoded signal for the data input sequence of "01100001010000." Assume the last signal level was negative.

10. Show graphically the HDB3 encoded signal for the data input sequence of "10100001110000." Assume the last signal level was negative.

Non Return to Zero (NRZ) Encoding Method

Introduction: Briefly describe the NRZ encoding method, its different types, its advantages, disadvantages, and applications. Also write a brief description of a pseudorandom number generator (PRNG).

Parts and Equipment

- 1 kΩ and 10 kΩ (3) resistors
- SPDT switch (2)
- LM 741 Operational amplifier
- 74LS86 XOR gate
- 74LS74 D-flip-flop
- 74LS164 8-bit parallel-out shift register
- LED (2), any color
- Two power supplies (5 V and 12 V-DC)
- Function generator
- Oscilloscope

1. Construct the NRZ encoding circuit as shown in Figure 4–27.

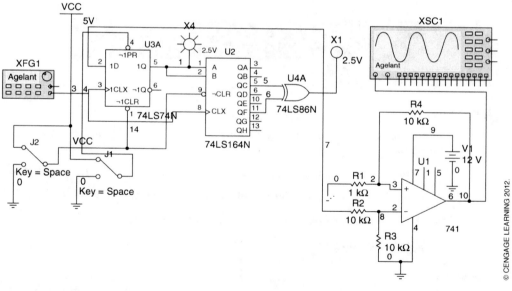

Figure 4–27: NRZ encoding circuit

2. Set the function generator at 10 V$_{p-p}$ and 50 Hz square wave.

Continues on next page

3. Turn the /PRE switch to OFF and then to ON to activate the PRNG.

4. Observe the NRZ signal, as shown in Figure 4–28, at the oscilloscope.

5. You may change the frequency of the function generator for better observation.

6. Save and print the NRZ signal from the oscilloscope screen and include it in your laboratory report.

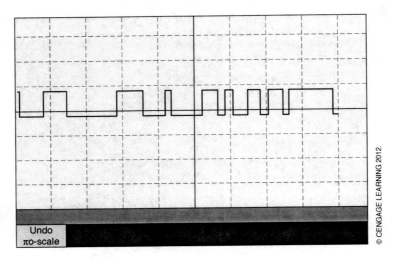

Figure 4–28: NRZ encoded signal

Questions

1. Why do you need to preset the output of the circuit at the beginning of this experiment?

2. What is the purpose of having an SPDT switch at the clear input of the D-flip-flop?

3. What is the function of the 8-bit parallel-out shift register in the NRZ circuit?

4. What is the function of the op amp circuit in the NRZ circuit?

5. What should you do to have greater output signal amplification?

Conclusion

Unipolar Return to Zero (RZ) Encoding Method

Introduction: Briefly describe the unipolar return to zero encoding method, its advantages and disadvantages in comparison to NRZ, and its applications.

Parts and Equipment

- 1 kΩ and 10 kΩ (3) resistors
- SPDT switch (2)
- LM 741 Operational amplifier
- 74LS08 AND gate
- 74LS86 XOR gate
- 74LS74 D-flip-flop
- 74LS164 8-bit parallel-out shift register
- LED (2), any color
- Two power supplies (5 V and 12 V-DC)
- Function generator
- Oscilloscope

1. Construct the unipolar RZ encoding circuit as shown in Figure 4–29.

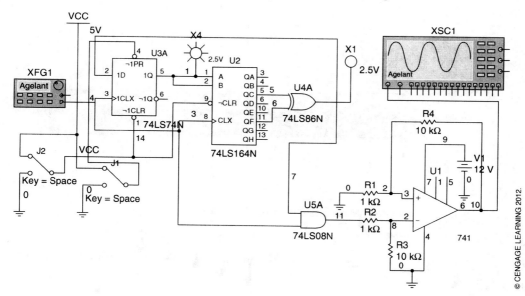

Figure 4–29: Unipolar RZ encoding circuit

© CENGAGE LEARNING 2012.

2. Set the function generator at 10 V$_{p\text{-}p}$ and 50 Hz square wave.

Continues on next page

3. Turn the /PRE switch to the OFF and then to the ON positions to activate the PRNG (pseudorandom number generator).

4. Observe the unipolar RZ signal, as shown in Figure 4–30, at the oscilloscope.

5. You may change the frequency of the function generator for better observation.

6. Save and print the unipolar RZ signal from the oscilloscope screen and include it in your laboratory report.

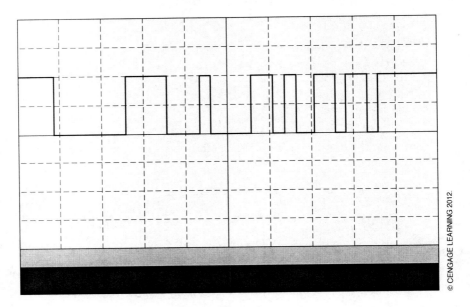

© CENGAGE LEARNING 2012.

Figure 4–30: Unipolar RZ encoded signal

Questions

1. What are the differences between NRZ and unipolar RZ circuitry?

2. What would happen if you turn OFF both SPDT switches?

3. What would happen if you turn OFF the preset switch and turn ON the clear switch?

4. What are the differences between unipolar and bipolar signals?

5. What are the advantages and disadvantages of unipolar RZ in comparison with the NRZ encoded signal in the first experiment?

Conclusion

Bipolar NRZ Encoding Method

Introduction: Briefly describe the bipolar non return to zero encoding method, its advantages and disadvantages in comparison with RZ, and its applications.

Parts and Equipment

- 1 kΩ and 10 kΩ (3) resistors
- SPDT switch (2)
- LM 741 Operational amplifier
- 74LS08 AND gate
- 74LS74 D-flip-flop
- 74LS86 XOR gate
- 74LS112 JK-flip-flop
- 74LS164 8-bit parallel-out shift register
- LED (2), any color
- Two power supplies (5 V and 12 V-DC)
- Function generator
- Oscilloscope

1. Construct the bipolar NRZ encoding circuit as shown in Figure 4–31.

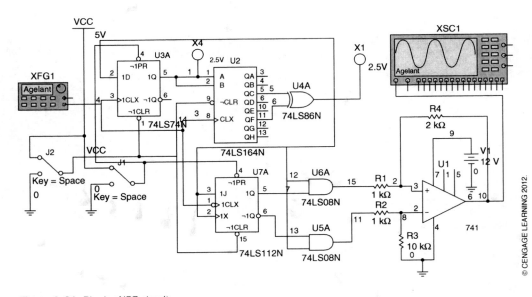

Figure 4–31: Bipolar NRZ circuit

2. Set the function generator at 10 V$_{\text{p-p}}$ and 50 Hz square wave.

3. Turn the /PRE switch to the OFF and then to the ON position to activate the PRNG.

4. Observe the bipolar NRZ signal, as shown in Figure 4–32, at the oscilloscope.

5. You may change the frequency of the function generator for better observation.

6. Save and print the bipolar NRZ signal from the oscilloscope screen and include it in your laboratory report.

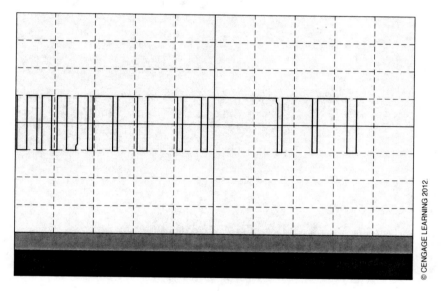

© CENGAGE LEARNING 2012.

Figure 4–32: Bipolar NRZ encoded signal

Questions

1. What are the differences between NRZ and bipolar NRZ circuitry?

2. What is the reason to have two AND gates in the bipolar NRZ circuit instead of having only one AND gate as in the unipolar RZ circuit?

3. What is the function of the JK flip-flop in this circuit?

4. What would happen if you replace AND gates with NAND gates?

5. Why does the output signal of the comparator not go to zero level?

Continues on next page

6. What are the advantages and disadvantages of bipolar NRZ in comparison with the NRZ encoded signal in the first experiment?

Conclusion

Differential Manchester Encoding Method

Introduction: Briefly describe the differential Manchester encoding method and its applications. Discuss its advantages and disadvantages in comparison with the Manchester encoding method.

Parts and Equipment

- 74LS00 NAND gate
- 74LS04 Inverter
- 74LS08 AND gate
- 74LS86 XOR gate
- 74LS74 D-flip-flop
- 74LS73 Dual JK flip-flop with no preset input
- 74LS164 8-bit parallel-out shift register
- LED (2), any color
- Two power supplies (5 V and 12 V-DC)
- Function generator
- Oscilloscope

1. Construct the differential Manchester encoding circuit as shown in Figure 4–33.

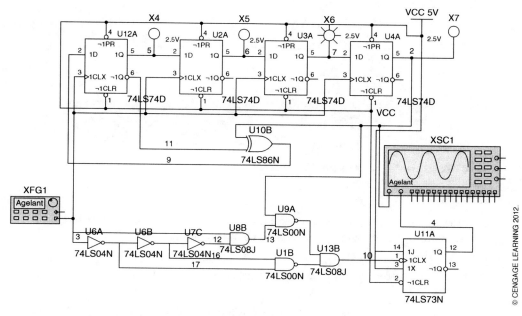

Figure 4–33: Differential Manchester encoding circuit

© CENGAGE LEARNING 2012.

Continues on next page

2. Set the function generator at 10 V$_{p-p}$ and 30 Hz square wave.

3. Connect the output of the PRNG to the first channel and the output of the JK flip-flop to the second channel of the oscilloscope.

4. Turn ON the function generator and observe the input data input signal and encoded differential Manchester signal, as shown in Figure 4–34, on the screen of the oscilloscope.

5. You may change the frequency of the function generator for better observation.

6. Save and print both the input and encoded differential Manchester signals from the oscilloscope screen and include them in your laboratory report.

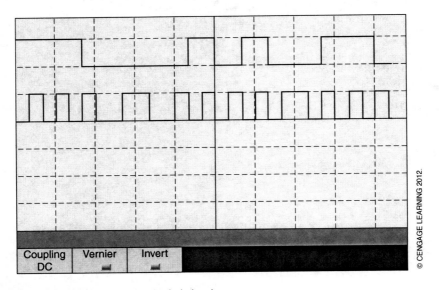

Figure 4–34: Differential Manchester encoded signal

Questions

1. Does the output signal of the second channel represent a perfect differential Manchester encoded signal? Why? Explain in detail.

2. What is inverse differential Manchester encoding? What are the differences between differential Manchester and inverse differential Manchester encoded signals?

3. Why do we need a series of inverters? What are their specific outputs?

4. Why do we need a JK flip-flop with clear input only (no preset input)? What is the function of the clear input?

Conclusion

CHAPTER

5

Issues in Data Communication

Objectives

After completing this chapter, students should be able to:

Discuss the fault analysis in a communication system.

Discuss the different methods of multiplexing in a data communication system.

Analyze the techniques of error detection and correction.

Describe encryption and data compression.

The encoding and transmission methods discussed in the last chapter are the processes that should be followed to prepare data for transmission. Preparation of data for transmission is only one aspect of data communication, and other parameters such as performance, efficiency, reliability, and error-free transmission need to be carefully considered in the design and construction of data communication. In this chapter we will discuss a variety of methods that can help increase the performance, efficiency, and accuracy of data communication in a network system.

The performance of a network system used for data communication depends on two major factors: (1) the way the network was designed, managed, and built; and (2) faults due to the improper operation of the transmission line and communication equipment.

5.1 REQUIREMENTS FOR A NETWORK THAT IS WELL-DESIGNED, WELL-MANAGED, AND WELL-BUILT

A well-designed, well-managed, and properly built network system results in reliability, recovery, security, and consistency. As a result, the network system will be dependable, will operate properly, and data can be sent and received in a timely manner with little to no error. The characteristics of a network system that performs well are defined below.

Reliability

When a network fails, users will often want to know how long it will take to reestablish the communication at the predetermined level of operation. Therefore, reliability of a data communication system is defined as the ability of the system to reestablish lost communication and to be operational at its specified level in a minimum amount of time following a loss of communication.

The specified level of operation of a device, piece of equipment, or system is posted by the manufacturer and is measured by the average time that a component needs to become operational following a failure. This average time is known as the mean time between failures (MTBF).

Recovery

There are always backup files for the data that need to be transmitted in a network system. The backup files help to recover the data lost during transmission. Recovery is defined as the ability of a system to reestablish the operation of the network with a minimum amount of data lost in comparison with the backup files.

Security

To protect a network system from any internal or external harm to the system hardware or software, a set of security features, such as antivirus protection, user passwords, encryption of data, access codes, and other means of security are usually implemented within a network system.

Consistency

A network system must operate with consistency in such features as response time, performance time, recovery time, and other related issues.

5.2 FAULTS DUE TO IMPROPER OPERATION OF THE TRANSMISSION LINE AND COMMUNICATION EQUIPMENT

The signal transmission over a physical channel, its surrounding environment, and the attached equipment on the transmission and receiver sides also affect the performance quality of a data communication system. In this section, some important factors, such as echo, crosstalk, pitch and frequency shifting, jitter, dropout packets, impulse noise, delay, bandwidth, bandwidth-delay product, and throughput will be defined.

Echo

If the transmitted signal is terminated or is being totally absorbed at the receiver side, it will be reflected back to the transmission side. This phenomenon is called *echo*. Echo is mostly an issue in voice transmission and specifically when a full duplex mode is being used. Elimination of echo is possible by using *echo cancellers*. Almost all modems use echo cancellers.

Crosstalk

Signal transmission in a wire is actually the movement of a current signal. The movement in a wire along with the applied electric field creates an electromagnetic field around the wire. If the wire is not shielded perfectly, electromagnetic wave interference (EMI) will occur between the waves in the wire and its adjacent wire. This phenomenon is called crosstalk. Crosstalk in voice communication will result in receiving more than one voice signal. Twisting wires and using two different twisted wires for the adjacent wires should eliminate the effect of crosstalk. Untwisted wires are the major source of crosstalk. Therefore, communication systems that use untwisted pair wires are more susceptible to crosstalk. Untwisted wires should always be kept very short to prevent crosstalk, even if it is necessary to use them in the patch panels and cable connectors. Near-end crosstalk (NEXT) occurs at the end of the cable carrying the signal (about 30 meters long from the transmission point) or when connectors are attached to twisted pair wires. Far-end crosstalk (FEXT) is a source of error in a short cable, since it will be attenuated and become very weak for a long cable. Thus, cables that are carrying high speed data (100 Mb or higher) must be tested for NEXT and FEXT.

Frequency and Pitch Shifting

In a communication system, frequency and pitch shifting result from local oscillator synchronization deficiency. This means the transmitter and receiver frequency are the same. This is often a concern in voice and music transmissions.

In frequency shifting, the fundamental and harmonic frequencies of the transmitted signal will change nonlinearly. For example, suppose the fundamental, first, and second harmonic frequencies are 1, 3, and 5 kHz, respectively. Their values will be shifted by an additional frequency (i.e., 2 kHz) and, therefore, their frequencies on the receiving end will be 3, 5, and 7 kHz, respectively. Consequently, the frequency ratio between them is the same and they lose their harmonic relationship.

In pitch shifting, however, the change in frequency is linear. This means the fundamental and harmonic frequencies at the receiver will be multiples of transmitter frequencies by a constant factor (i.e., 2 kHz). Therefore, frequencies at the receiver side for the same example will be 3, 6, and 10 kHz, respectively. As you may notice, the ratio of frequencies remains the same and they will not lose their harmonic relationship.

Jitter

Although many definitions exist for jitter in the field of communication, the most widely accepted is: "The small variation of a significant instant of a digital signal from its ideal position in time." In short, jitter is a timing error within a network system. If this definition is accepted, then jitter is not the same as latency in data communication but it still causes the information packets to be received at different time intervals at the receiver side. Jitter is used to measure latency of the information packet over time in a network system. Jitter is not actually noise but it is a source of data error that is called the bit error rate (BER). The existence of jitter, even at a low level, is an important issue in real-time applications, specifically in the case of voice and video communications. In a network system, jitter buffers are used to detect jitter.

According to the above definition, the "small variation" can happen in signal amplitude, phase, or frequency. This leads to three different types of jitter. Amplitude jitter is mostly caused by power supply noise and a ringing tone on the signal. Variation in phase may cause one pulse to move to the time allocation that is assigned for its adjacent pulse. Variation in frequency may result in latency in arrival time.

There are two additional types of jitter: random (unbounded) and deterministic (bounded). Random jitter is unpredictable jitter with no upper bound value and usually follows a Gaussian distribution similar to white noise. To keep the performance of a system at the optimal level, random jitter must be far less than 1 nsec. Deterministic jitter is predictable and repetitive jitter in the clock recover circuits of a network system. Deterministic jitter has a bounded peak-to-peak value and comes from many sources, such as intersymbol interference, duty-cycle distortion, and switching power supply.

Dropout Packets

To increase the reliability of industrial control systems, a communication network is placed in its feedback path instead of using simple point-to-point connections. This type of system is known as a *network control system* (NCS). Besides transmitted time delays, dropout in packets is an important issue in an NCS. The dropout packet phenomenon in an NCS is mainly due to the limited bandwidth of the system and affects the analysis, synthesis, and reliability of the system.

Impulse Noise

Impulse noise or spike is a transient sharp noise with very short duration time and random amplitude. Impulse noise is one of the most damaging factors in data communication and specifically in Asymmetric Digital Subscriber Line (ADSL) systems. A strong impulse noise can interfere with a telephone line if it is installed close to a dimmer switch, an electric AC power switch, or similar devices. Impulse noise will distribute evenly over the transmitting system. Impulse noise creates more problems when its repetition rate is high and its peak power is stronger than the background noise power.

Delay

Even if the internal processing time for transmission and reception of information such as frame formatting can be completely ignored, it will take time, called *delay* or *latency*, to transmit data from one end to the other. In general, the communication equipment (transmitter and receiver), transmission line, or intermediate devices can cause delays in the system.

Information packets consist of several hundred bits of data and it takes time to format them for proper transmission. Information leaves a transmitter bit by bit and there is a time delay equal to the time between receiving the first bit and the last bit. This type of time delay is directly proportional to the size of the information packet and inversely proportional to the data rate (bandwidth) of the digital system.

$$\text{Time delay (system)} = \frac{\text{information packet size (in bits)}}{\text{data rate (in bps)}} \tag{5.1}$$

Example 1: A 5.6-Mb information packet is ready to be transmitted by a digital communication system having a data rate of 2.2 Gbps. Find the total receiving time from the first to the last bit.

Solution:

$$\text{Time delay (system)} = \frac{(5.6)\,(8)\,(10^6)}{(2.2)\,(10^9)} = 2.36 \text{ msec}$$

Delay in the transmission line, or propagation time, is due to the fact that the Fourier components of an electromagnetic wave signal travel in the transmission line at different speeds. An electromagnetic wave travels with the speed of light (c) in the air (vacuum) but will decrease as the reflective index (n) of the medium (transmission line) increases. For example the speed of light in a copper wire is about 1.234×10^5 km/sec. Therefore, the propagation time is inversely proportional to the speed of light (300,000 km/sec) in the medium, and obviously directly proportional to the distance of transmission.

$$v = \frac{c}{n} \tag{5.2}$$

$$\text{Propagation time} = \frac{\text{distance}}{\text{speed}} \tag{5.3}$$

Example 2: Calculate the propagation time in a glass optical fiber and a copper cable that are used for cross-country (3000 miles) communication.

Solution: The speed of the signal in a medium is the ratio of the speed of light in the air and the reflective index of the medium. The reflective index of the fused glass and copper are 1.459 and 2.43, respectively.

$$v \text{ (in the glass optical fiber)} = \frac{300,000 \text{ km/sec}}{1.459} = 2.056 \times 10^5 \text{ km/sec}$$

$$v \text{ (in the copper cable)} = \frac{300,000 \text{ km/sec}}{2.43} = 1.234 \times 10^5 \text{ km/sec}$$

$$\text{Propagation time (in the glass optical fiber)} = \frac{\text{distance}}{\text{speed}} = \frac{(3000)(1.6) = 4.8 \text{ km}}{2.056 \times 10^5 \text{ km/sec}}$$
$$= 23.34 \text{ μsec}$$

$$\text{Propagation time (in the copper cable)} = \frac{\text{distance}}{\text{speed}} = \frac{(3000)(1.6) = 4.8 \text{ km}}{1.234 \times 10^5 \text{ km/sec}} = 38.89 \text{ μsec}$$

Intermediate networking devices such as a router, hub, or switches are also the source of latency or delay. It takes time for a packet of information to fully enter and leave any intermediate networking device. This delay time depends on how fast, for example, a router receives and reroutes the packet. Congestion in communication, error detection, and correction are also determining factors in receiving and processing the information packet at the receiver.

Bandwidth

Bandwidth refers to bands of the frequency of sine wave or cosine wave signals (Fourier series of a periodic function) that can propagate over a channel with an acceptable attenuation to be able to recover the transmitted signal. In an analog channel, bandwidth determines the capacity of the channel and is measured in Hertz. For example, the bandwidth of coaxial cable that is used in telephone lines is 300 MHz. A channel with higher bandwidth has larger capacity. In data communication, binary digits (1 or 0) are transmitted as pulse signals. The number of these pulse signals that can be fitted in a limited bandwidth for transmission depends on their pulse width. If the pulse width decreases, more pulse signal can be transmitted. Higher pulse transmission means higher bit transmission or higher bit rate. This is the reason bandwidth in data communication is defined as *data rate*. Limitations of communication circuits and systems are imposed on bandwidth limitation.

Bandwidth-Delay Product

Bandwidth-delay product (BDP) is the product of the maximum available bandwidth in a link and the delay (round-trip time or RTT). BDP determines the number of data bits that can be fitted in a link for transformation before going to a stop-and-wait process or receiving packet acknowledgments. In a window-based protocol such as Transmission Control Protocol (TCP), however, this task is done by the TCP window and BDP is a tool that helps to determine the size of the receive window (RWIN). Therefore, BDP is an important parameter in a window-based protocol in order to fill the link and adjust the systems to respond to the specific type of network that is being used. TCP is covered in Chapter 10.

Throughput

Throughput is the amount of data or data packets that can be transferred successfully over a communicational channel or a network node per unit of time or time slot. Throughput is measured in bits per second, and usually stated in megabits per second (Mbps) or gigabits per second (Gbps) for today's super-fast data transfer systems. Even though throughput has the same unit as bandwidth in data communication, they do not use the same parameters. Bandwidth defines the potential data rate, while throughput defines the successful data rate in a data communication system. Throughput is bound by the bandwidth-delay product (BDP).

Example 3: Find throughput in a local area network that is capable of transferring 6000 packets of information in 20 seconds. Assume each packet of information contains 1500 bytes.

Solution:

$$\text{Throughput} = \frac{(6000)(1500)(8)}{20} = 3.6 \text{ Mbps}$$

Bandwidth Efficiency

Bandwidth efficiency is a parameter that is used for comparing the performance of one modulation method over another. It is sometimes called *information density* or *spectral efficiency*. Bandwidth efficiency is measured by the ratio of net transmission bit rate (excluding the error correction bits) to the minimum required bandwidth (useful bandwidth) for a particular modulation method. The normal value of the bandwidth efficiency is based on 1 unit of bandwidth which is 1 Hz.

Example 4: Find the bandwidth efficiency of a 56-K modem that sends a modulated signal over a telephone line.

Solution: The bandwidth of a telephone line is 3.1 kHz and the bit rate of a 56-K modem is 56 Kbps, therefore the bandwidth efficiency of the 56-K modem is:

$$\text{Bandwidth efficiency} = \frac{56 \text{ Kbps}}{3.1 \text{ kHz}} = 18.06 \text{ bits/sec/Hz}$$

5.3 MULTIPLEXING

Multiplexing or multichanneling is a technique for more efficient use of transmission media in data communication, which acts like actual data compression. In the multiplexing technique, multiple transmitters share one single channel to transmit their frequencies, data frames, or wavelengths. Multiplexing frequencies, time (time of data frames), and wavelengths are called frequency-division-multiplexing (FDM), time-division-multiplexing (TDM), and wavelength-division-multiplexing (WDM), respectively.

Frequency-Division-Multiplexing (FDM)

FDM is used to multiplex multiple low bandwidth analog signals simultaneously in order to transmit them in a single high bandwidth channel. Each transmitter signal

$[E_1(t), E_2(t) \dots E_n(t)]$ is modulated with predetermined local carrier signals to convert them to modulated signals of $[E_{m1}(t), E_{m2}(t) \dots E_{mn}(t)]$ before transmission. This will help avoid overlap and interference in their frequency domains. All signals have the same bandwidth size with different frequency ranges. For example, in the North American analog signal hierarchy, 12 signals with 4-kHz bandwidth are multiplexed to generate the 48-kHz voice channel. The band that is separated by the subcarrier frequencies is called the *guard band*. Modulated signals will be separated first by bandpass filters and then demodulated at the receiver side to recover the original signals. Cable TV is a well-known application of FDM, where multiple preassigned TV station signals with a 6-MHz bandwidth for each are multiplexed and transmitted over the cable TV. A variety of modulation techniques such as quadrature-phase shift keying (QPSK) and quadrature amplitude modulation (QAM) are used to encode data and transmit them over higher bandwidth channels using FDM. Figure 5–1 shows the process of FDM.

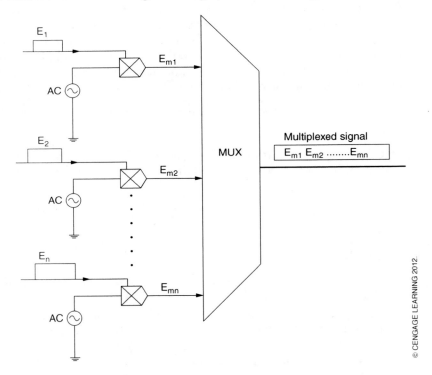

Figure 5–1: The frequency-division-multiplexing process

FDM is also used to multiplex lower bandwidth telephone lines to a higher bandwidth telephone line for increased efficiency. The North American FDM carrier standard (sometimes called the *North American hierarchical telephone line system*) is different than international standards. Its multiplexing systems and related bandwidth are divided into four different categories: Groups, Super Groups, Master Groups, and Jumbo Groups which are listed in Table 5–1.

Category	Number of voice channels that are multiplexed	Bandwidth	Spectrum
Groups	12	48 kHz	60–108 kHz
Super Groups	5	240 kHz	312–552 kHz
Master Groups	10	2.52 MHz	564–3084 kHz
Jumbo Groups	6	16.984 MHz	0.564–17.548 MHz

© CENGAGE LEARNING 2012.

Table 5–1: North American hierarchical FDM carrier standards

Time-Division-Multiplexing (TDM)

TDM is used to transmit several low data rate digital signals to a high data rate digital transmission medium (link or channel) at the same time and provide a better quality of service (QoS). The output signals of TDM can be also transmitted over an analog transmission medium but they have to go through a modem before being transmitted. TDM is divided into two categories: synchronous and statistical.

Synchronous Time-Division-Multiplexing

This method divides the bandwidth of a high data rate link into a series of equal time slots where each specific time slot is assigned to only one of the low rate sources. Therefore, each source has a fixed time slot in the link for transmission. Time slots are not necessarily divided evenly (this means multiple time slots can be assigned to a source that has more data to send). Data units can be either in bits, bytes, or frames, but in general it is organized into frames. Each data unit from the low rate will pass through buffers and then is scanned in a round-robin fashion before the multiplexing step. All sources should have the same data rate and the data rate of the link must be the product of the sources' data rate times the number of sources in the system. For example, if there are five sources of 10 Kbps data rate, then the bandwidth of the link must be divided into five equal time slots with a total data rate of 50 Kbps.

The advantage of synchronous TDM is that including the receiver address in the data frame is not required because of the synchronization between the multiplexer and demultiplexer, and the fixed time slot assignment for each source. The disadvantage is that the link capacity in synchronous TDM will not be used efficiently if there are one or more empty time slots. This occurs if one or more source(s) do not have any data to send. Statistical TDM was developed to resolve this problem and use the bandwidth of the link efficiently. Three types of synchronous TDM are T1 carrier, ISDN, and SONET. Figure 5–2 shows the process of synchronous TDM.

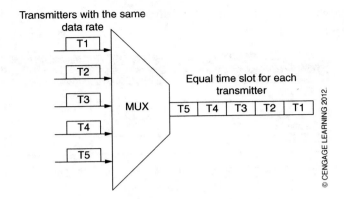

Figure 5–2: The time-division-multiplexing process

If the data rate is not the same for all sources, then they need to be adjusted in order to have the same data rate. Bit padding (or bit stuffing), a secondary multiplexer, and a secondary demultiplexer may be used to equalize the data rate of all sources.

Bit Padding Bit padding is used when one source has a lower data rate than other sources but the ratio of the higher to the lower data rate is not an integer. In this case some non-data bits (or dummy bits) are inserted into the frame of the lower data rate source to equalize its data rate with others. For example, if one out of five sources of a multiplexer has 126 Kbps and other sources have 128 Kbps data rates, then 2000 recognizable non-data bits will be inserted into the 120-Kbps source frame and will be disregarded at the demultiplexer side. Figure 5–3 shows a block diagram of the bit padding process.

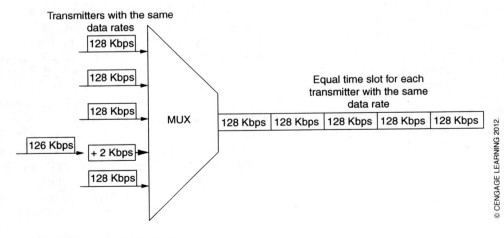

Figure 5–3: The bit padding process

Secondary Multiplexer A secondary multiplexer is used when two or more sources have lower data rates but the ratio of the higher to the lower data rate is an integer. In this case, sources with lower data rates are multiplexed with each other to equal the data rate of all the sources that are attached to the main multiplexer. A secondary de-multiplexer is needed to convert the higher to the lower data rate. For example, if two out of five sources of a multiplexer have a data rate of 64 Kbps and other sources have data rates of 128 Kbps, then a secondary input multiplexer will be used for the sources with lower data rate. Figure 5–4 shows the function of the secondary multiplexer.

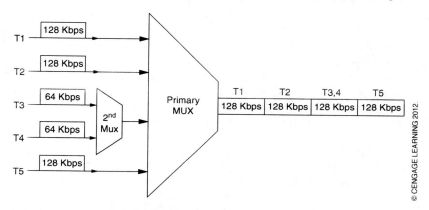

Figure 5–4: The function of the secondary multiplexer

Secondary Demultiplexer A secondary demultiplexer is used when one source has a higher data rate than others but the ratio of the higher and lower data rate is an integer. For example, if one source has a data rate of 128 Kbps and others have 64 Kbps, then a one-to-two demultiplexer may be used to divide the 128 Kbps to two 64 Kbps, so that all sources will have a data rate of 64 Kbps. A two-to-one multiplexer at the demultiplexer will convert the two lower rates back to the higher data rate signal. A serial-to-parallel converter may also be used instead of a secondary demultiplexer. Figure 5–5 shows the function of the secondary demultiplexer.

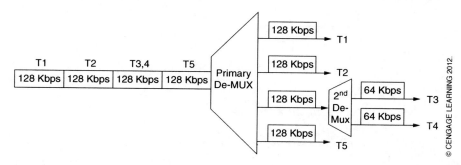

Figure 5–5: The function of the secondary demultiplexer

Statistical Time-Division-Multiplexing

Statistical TDM is an intelligent TDM and assigns a time slot only for an active source. This means there are no fixed time slots assigned for any source. An active source is a source that has data to send. In statistical TDM, the number of time slots in the link frame is variable and depends on how many sources have data to send. Variable data size is also permitted.

In statistical TDM, the statistical average transmission rate of the input data streams that are to be multiplexed is calculated, and then a multiplexing link that has a data rate equal to or higher than the statistical average transmission rate is selected. Unlike synchronized TDM, there is no need to assign a fixed number of time slots to each input data stream. To ensure the transmitted information will arrive at the correct destination, a header is added to each block of the input data stream. Each header consists of the start and end of the block of data, the address field, control field, error control, and the information field. In statistical TDM, a buffer is used to prevent the data rate of the multiplexed data block from exceeding the transmission rate. Table 5–2 shows the format of the statistical TDM data block.

Start flag	Address field	Control field	Information field	Error control	End flag

Table 5–2: The format of the statistical TDM data block © CENGAGE LEARNING 2012.

An advantage of statistical TDM is that it makes efficient use of the link capacity. It is a good choice for a low bandwidth link. A disadvantage is that the receiver's address must be included in the time slot since there is no fixed time slot assigned. Figure 5–6 shows the process of statistical TDM.

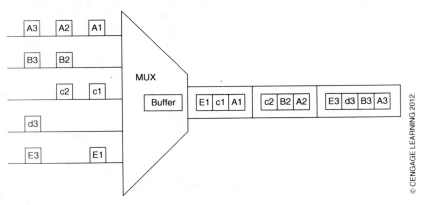

Figure 5–6: The statistical TDM process

Transmission rate standards in North America are also different from European standards. Digital signal level zero (DS-0) is the standard base transmission rate in North America and, as was discussed briefly in Chapter 2, the rate for DS-0 is 64

Kbps based on the Nyquist theorem. According to this standard, one voice channel has a data rate of 64 Kbps. The unipolar encoding method is used in a DS-0 signal. The North American hierarchy of TDM carrier standards is shown in Table 5–3.

Carrier system	Digital signal level (frame format)	No. of voice channels	Bit rate (Mbps)	Encoding methods	Media
T–1	DS–1	24	1.544	Bipolar RZ	22 AWG
T–1C	DS–1C	48	3.152	Bipolar RZ	22 AWG
T–2	DS–2	96	6.312	Bipolar RZ and B6ZS	22 AWG
T–3	DS–3	672	44.736	RZ and B3ZS	Microwave and optical fiber
T–4	DS–4	4032	274.176	Bipolar NRZ	Coaxial cable and microwave

Table 5–3: The North American hierarchy of TDM carrier standards

DS-1 Frame Format

The DS-1 frame format includes 24 voice channels (24 DS-0) plus 1 framing bit. Each channel has 8 bits and therefore, the DS-1 frame has $24 \times 8 = 192$ bits + 1 framing bit = 193 bits. Since the sampling frequency for DS-0 is 8000 Hz, the data rate for DS-1 is 193 bits $\times$ 8000 Hz = 1,544,000 bit/sec or 1.544 Mbps. The frame duration is 125 µsec, with 0.6477 µsec for each bit and 5.18 µsec for each channel. The framing bit is used for synchronization and alignment. A group of 12 frames is called a multiframe and each frame begins with a start bit (S-bit). S-bits identify the start of each frame. Their bit values are shown in Table 5–4.

Frame	1	2	3	4	5	6	7	8	9	10	11	12
S-bit	**1**	0	**0**	0	**1**	1	**0**	1	**1**	1	**0**	0

Table 5–4: S-bits in the DS-1 frame

The odd-numbered frames have a pattern of 101010 (shown in bold in Table 5–3) and are used for frame synchronization. The even-numbered frames have a pattern of 001110 and are used for multiframe synchronization.

DS-1 is used in T-1 lines and if T-1 is used for data only, then channel 24 is used for special synchronization that would help with faster service and reliable framing. The DS-1 frame format is shown in Table 5–5.

DS-1 frame duration = 125 µsec frame																							
24 DS-0 information with 8 bits each = 192 bits																					Framing bit		
1	2	3	4	5	6	7	8	9	10	11	12	13	14	15	16	17	18	19	20	21	22	23	24

Table 5–5: The DS-1 frame format

DS-2 Frame Format

The DS-2 frame format includes four DS-1 signals ($4 \times 193 = 772$ bits) plus overhead and stuffing bits = 789 bits with data rate of 789 bits $\times$ 8000 Hz = 6.312 Mbps. The DS-2 frame consists of four DS-1 subframes of M1, M2, M3, and M4. Each subframe has six blocks and each block has 49 bits each with a total of $6 \times 49 = 294$ bits per subframe. Each block starts with an overhead bit (OH-bit) and follows with 48 bits of DS-1 information bits. Each DS-2 frame has 24 overhead bits (4 subframes $\times$ 6 blocks $\times$ 1 OH-bit = 24). There are three different types of overhead: frame bits (F-bits), multiframe bits (M-bits), and control bits (C-bits).

F-bits or framing bits are referencing bits of demultiplexing of the four DS-1 channels. F-bits are located in blocks 3 and 6 of each subframe with a sequence of "01" where level "0" is assigned to block 3 and level "1" is assigned to block 6.

M-bits are used to identify the subframes. M-bits are located only in the first block of each subframe; therefore, there are only four M-bits in the DS-2 frame with a sequence of 011X, where level "0" is assigned to the first subframe, level "1" to the second and third subframes, and level "X" (X can be either 0 or 1) assigned to the fourth subframe.

C-bits are used to transport information on the stuffing bits. Stuffing bits are used to synchronize the four DS-1 signals. There are three C-bits per subframe and

M1 Frame											
Block 1		Block 2		Block 3		Block 4		Block 5		Block 6	
OH-bit	48 DS-1 info bits	OH-bit	48 DS-1 info bits	OH-bit	48 DS-1 info bits	OH-bit	48 DS-1 info bits	OH-bit	48 DS-1 info bits	OH-bit	48 DS-1 info bits
M-bit = 0		C1-bit		F-bit = 0		C2-bit		C3-bit		F-bit = 1	
M2 Frame											
Block 1		Block 2		Block 3		Block 4		Block 5		Block 6	
OH-bit	48 DS-1 info bits	OH-bit	48 DS-1 info bits	OH-bit	48 DS-1 info bits	OH-bit	48-DS-1 info bits	OH-bit	48 DS-1 info bits	OH-bit	48 DS-1 info bits
M-bit = 1		C1-bit		F-bit = 0		C2-bit		C3-bit		F-bit = 1	
M3 Frame											
Block 1		Block 2		Block 3		Block 4		Block 5		Block 6	
OH-bit	48 DS-1 info bits	OH-bit	48 DS-1 info bits	OH-bit	48 DS-1 info bits	OH-bit	48 DS-1 info bits	OH	48 DS-1 info bits	OH-bit	48 DS-1 info bits
M-bit =1		C1-bit		F-bit = 0		C2-bit		C3-bit		F-bit = 1	
M4 Frame											
Block 1		Block 2		Block 3		Block 4		Block 5		Block 6	
OH-bit	48 DS-1 info bits	OH-bit	48 DS-1 info bits	OH-bit	48 DS-1 info bits	OH-bit	48 DS-1 info bits	OH-bit	48 DS-1 info bits	OH-bit	48 DS-1 info bits
M-bit = x		C1-bit		F-bit = 0		C2-bit		C3-bit		F-bit = 1	

Table 5–6: The DS-2 M-frame format

a total of 12 C-bits in the DS-2 frame. If all three C-bits are "0" then there is no stuffing and if all C-bits are "1" this means the C-bits shown are stuffing. C-bits are located in blocks 2, 4, and 5. Table 5–6 shows the DS-2 M-frame format.

Wavelength-Division-Multiplexing

Wavelength-division-multiplexing (WDM) is used in optical communication where the transmitted signals are optical signals (lights). Light signals are characterized by their wavelength, which is shown by λ and usually measured in nanometers (nm). The wavelength of a light source indicates its color and it is inversely proportional to its frequency, $\lambda = c/f$ where c is the speed of light. Therefore, WDM, which is also called *dense wavelength-division-multiplexing* (DWDM), is similar to FDM but uses optical signals instead of electric signals. In WDM several discrete wavelengths are combined and transmitted over an optical fiber. Each wavelength is actually one independent channel with a data rate from 2.5 Gbps to over 100 Gbps. This means optical communication is much faster than that using copper cable, and data can be transferred at a rate of several hundred gigabits per second. The wavelength of light that is commonly used in WDM is about 1550 nm in response to the capability of an erbium-doped fiber amplifier (EDFA). Figure 5–7 shows the process of WDM.

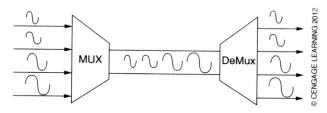

© CENGAGE LEARNING 2012

Figure 5–7: The wavelength-division-multiplexing process

5.4 ERROR DETECTION AND CORRECTION

Data may have been degraded or corrupted in one or multiple bits during the transmission period; therefore, when it arrives at the receiver side, it will not be the same as at its origin. To receive the correct data at the receiver side, many different techniques have been developed. In this section the most common techniques will be discussed in detail.

Parity Generator and Checker

A parity generator and checker is an electronic circuit that is used to detect a single-bit error in the data transmission. As was discussed in Chapter 4, in the asynchronous serial transmission method a parity bit will be generated by the parity generator and appended to the data bits. This will help detect any possible single-bit errors that may occur during transmission. A parity generator will employ $n-1$ XOR gates to generate a parity bit for n-bit data at the transmission side. On the receiver side, a parity checker will repeat the same procedure that the parity generator does and will compare the two parity bits. There is no error in the data transmission if the two parity bits are identical. Therefore, to determine if there is an error during transmission, the parity checker uses one more XOR gate than the parity generator.

The parity generator and checker are either even or odd. The parity bit in an even parity bit generator is 1 if there is an odd number of 1 bits in the data bits. The parity checker will produce bit 1 if there is an even number of 1 bits in the total number of data bits of the parity checker (including the parity bit). This means an error occurred in the data bits during transmission. The odd parity generator and checker performs the opposite procedure as the even parity generator and checker. The parity checker is able to detect a single-bit error in the data transmission but will not be able to correct it. It is a very effective method with white noise but not with noise bursts. Figure 5–8 shows a 4-bit parity generator and checker circuit.

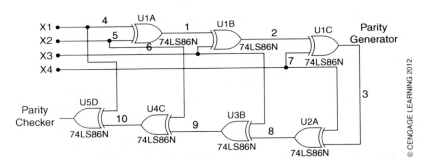

Figure 5–8: A 4-bit parity generator and checker

Example 5: Determine both even and odd parity bits for transmission of the letter T and show the transmitted data bit frame in the asynchronous serial transmission method.

Solution: There are three 1 bits in the American Standard Code for Information Interchange (ASCII) code representation of the letter T, which is "1010100." Therefore, the even parity bit will be a 1 bit and the odd parity bit will be a 0 bit.

> The even parity bit = 1 and the transmitted data bit frame will be "10101001"
>
> The odd parity bit = 0 and the transmitted data bit frame will be "10101000"

Longitudinal and Vertical Redundancy Check

Longitudinal redundancy check (LRC) and vertical redundancy check (VRC) are improved versions of the parity method that are able to detect and correct multiple bit errors without requesting any additional information from the transmitter because of their two-dimensional parity check nature. This process of correcting error bits is known as *forward error correction* (FEC). The procedure of the LRC and VRC is as follows:

1. Write the information that is set for transmission character by character in a row.
2. Write their ASCII code representations under each individual character starting from the least significant bit (LSB) to the most significant bit (MSB).
3. Apply either odd or even parity to determine the parity bit for each row and write it down in the block check character (BCC) column.
4. The BCC column that has the same number of bits as other characters have will be transmitted as a character at the end of the information block.
5. The receiver will follow the same procedure as above and generate its own BCC and will compare it with the transmitter's BCC. The same parity method must be used to find the LRC and VRC bits. If both BCCs are the same, then there is no error in the transmitted information. If there is an error, the LRC and VRC bits will be distorted and it will help to locate and correct the error bit.

LRC along with VRC can always correct a single-bit error but not always multiple bit errors. The following examples show the limits of this method.

Figure 5–9 shows the electronic circuit to generate vertical parity bits. The bias is used to generate an odd parity (bias bit is 1) or even parity (bias bit is 0).

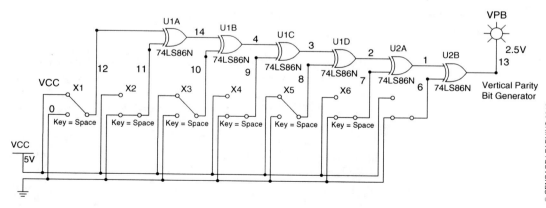

Figure 5–9: Vertical parity bit generator

© CENGAGE LEARNING 2012.

Example 6: The word "Network" is set to be transmitted. Determine the following:

1. The LRC, VRC, and BCC bits using odd parity for both LRC and VRC.
2. How the LRC and VRC methods detect a single-bit error. Assume bit four (X4) of the character "t" has been altered during transmission.
3. How the LRC and VRC methods detect double-bit errors. Assume bits four (X4) of the character "t" and character "o" have been altered during transmission.
4. How the LRC and VRC methods detect a single-bit error. Assume bit four (X4) of the character "t" and bit five (X5) of character "o" have been altered during transmission.
5. How the LRC and VRC methods detect a four bit error. Assume bit four (X4) and bit five (X5) of the characters "t" and "w" have been altered during transmission.

Solution: a. Use the ASCII code shown in Table 5-7 to find the 7-bit code for each character of the word "Network" and place them in the table.

	ASCII Code	N	e	t	w	o	r	k	BCC bits
VRC	X1	0	1	0	1	1	0	1	1
	X2	1	0	0	1	1	1	1	0
	X3	1	1	1	1	1	0	0	0
	X4	1	0	0	0	1	0	1	0
	X5	0	0	1	1	0	1	0	0
	X6	0	1	1	1	1	1	1	1
	X7	1	1	1	1	1	1	1	0
Parity bit		1	1	1	1	1	1	0	**1**
			L	R	C				

Table 5–7: VRC and LRC bits of the word "Network"

LRC is the combination of the parity bit for each character and varies according to the number of characters in the information. In this example, LRC is an 8-bit binary number of "11111101." VRC depends on the 7-bit ASCII code of each character in the information block. The BCC word is a 7-bit word like all other characters in the information block.

In this example, the 7-bit representation of the BCC word is "1000010". You may have noticed that the bolded bit 1 in the last block in the BCC column, which is called the *BCC parity bit,* satisfies the odd parity of both the LRC and BBC.

1. In this example, the change in bit four of letter "t" from "0" to "1" will cause an alteration of bit three in the LRC row and bit four in the BCC column at the receiver's side. A change in LRC will notify the receiver

about the error that has occurred in the character "t" and the change in BBC will help the receiver to know that the error happened in bit four of the character "t." With this information, the receiver will perform the FEC process to correct the error. Compare Table 5–7 and Table 5–8 to realize the bit error detection and correction in this example. All altered bits are shown in bold.

	ASCII Code	N	e	t	w	o	r	k	BCC bits
V R C	X1	0	1	0	1	1	0	1	1
	X2	1	0	0	1	1	1	1	0
	X3	1	1	1	1	1	0	0	0
	X4	1	0	**1**	0	1	0	1	**1**
	X5	0	0	1	1	0	1	0	0
	X6	0	1	1	1	1	1	1	1
	X7	1	1	1	1	1	1	1	0
Parity bit		1	1	**0**	1	1	1	0	**0**
			L	R	C				

© CENGAGE LEARNING 2012.

Table 5–8: Altered bits in character "t," LRC, and BBC

2. In this example, bit four of characters "t" and "o" have changed from "0" to "1" and from "1" to "0," causing changes in the third and fifth columns of LRC but no change in the BCC column. As a result, the receiver will detect a double-bit error but will not be able to perform FEC to correct the error. Compare Table 5–7 and Table 5–9 to realize the bit error detection but not correction in this example. All altered bits are shown in bold.

	ASCII Code	N	e	t	w	o	r	k	BCC bits
V R C	X1	0	1	0	1	1	0	1	1
	X2	1	0	0	1	1	1	1	0
	X3	1	1	1	1	1	0	0	0
	X4	1	0	**1**	0	**0**	0	1	0
	X5	0	0	1	1	0	1	0	0
	X6	0	1	1	1	1	1	1	1
	X7	1	1	1	1	1	1	1	0
Parity bit		1	1	**0**	1	**0**	1	0	1
			L	R	C				

© CENGAGE LEARNING 2012.

Table 5–9: Altered bit in the characters "t," "o," and LRC

3. In this example, bit four of character "t" has changed from "0" to "1" and bit five of character "o" has changed from "1" to "0." These changes in bits will result in alterations of the third and fourth bits in the LRC, and the fourth and fifth bits of the BCC. As a result, the receiver will find two errors in its error detection and correction device: one error in the fourth row and third column (the fourth bit of the character "t") and another error in the fifth row and fifth column (the fifth bit of the character "o"). With this information, the receiver would able to correct both errors by performing the FEC process. Compare Table 5–7 and Table 5–10 to realize the bit error detection and correction in this example. All altered bits are shown in bold.

	ASCII Code	N	e	t	w	o	r	k	BCC bits
V R C	X1	0	1	0	1	1	0	1	1
	X2	1	0	0	1	1	1	1	0
	X3	1	1	1	1	1	0	0	0
	X4	1	0	**1**	0	1	0	1	**1**
	X5	0	0	1	1	**1**	1	0	**1**
	X6	0	1	1	1	1	1	1	1
	X7	1	1	1	1	1	1	1	0
Parity bit		1	1	**0**	1	**0**	1	0	1
			L	R	C				

Table 5–10: Altered bits in the characters "t," "o," LRC, and BBC

4. In this example, bit four and bit five of characters "t" and "w" have changed from "0" to "1" and from "1" to "0," respectively. These errors in the transmitted bits will not affect the LRC and BBC bits and

	ASCII Code	N	e	t	w	o	r	k	BCC bits
V R C	X1	0	1	0	1	1	0	1	1
	X2	1	0	0	1	1	1	1	0
	X3	1	1	1	1	1	0	0	0
	X4	1	0	**1**	**1**	1	0	1	0
	X5	0	0	**0**	**0**	0	1	0	0
	X6	0	1	1	1	1	1	1	1
	X7	1	1	1	1	1	1	1	0
Parity bit		1	1	1	1	1	1	0	1
			L	R	C				

Table 5–11: Altered bits in the characters "t" and "o"

as a result the receiver will not be able to detect or correct any error bits. This is the worst scenario in this method of error detection and correction. Compare Table 5–7 and Table 5–11 to realize the bit error detection but not correction in this example. All altered bits are shown in bold.

Checksum

Checksum is the basic method of checking errors that was developed based on the redundancy of bits in the data and has been used in XMODEM and Internet Protocol (IP). Checksum is an error detection method in which the numerical value of characters in a message will be added together to generate a checksum character that will be appended to the end of the message before transmission. At the receiver's side, the same process of addition will take place to determine the receiver's checksum character. The receiver will compare its own checksum character with the transmitter's checksum character. If they are the same, then there is no error in the received message. If they are not the same, then an error has occurred. The easiest checksum processes is just adding the ASCII code of each character in the message and using the least significant byte of the result as a checksum character.

Example 7: Determine the checksum character of the word "Network."

Solution: Use the ASCII table to find the hexadecimal numerical value of each character in the word "Network" and then add them up.

$$
\begin{array}{ll}
\text{N} & \text{4E} \\
\text{e} & \text{65} \\
\text{t} & \text{74} \\
\text{w} & \text{77} \\
\text{o} & \text{6F} \\
\text{r} & \text{72} \\
\text{k} & \text{6B +} \\
\hline
 & \textbf{02EA}
\end{array}
$$

The checksum character would be EA, which is the same as the least significant byte in the addition.

The checksum character that is used in XMODEM is determined by taking 128 bytes of data, adding them up and dividing the result by 128.

In IP, the checksum is determined very differently than for XMODEM. At the transmission side, the checksum character will be set to 0 and then the message will be divided into 2-byte words (2-character words when the ASCII code is used). The 2-byte words are added together and the 16th complement of the result would be the checksum character. At the receiver, the 2-character words will be added again and the result is added to the checksum character. There is no error in the received message if the result of the final addition is 0.

Example 8: Determine the checksum character of the word "Network" for IP. Change the third bit of character "w" to show how checksum will detect the error.

Solution: Divide the word "Network" into groups of two characters. Add the numerical value of the two characters and find the 16th complement to find the checksum character. Add the checksum character to the result of the 2 character addition to find out if there is any error. Table 5–12 shows the process of checksum character generation and error detection.

At the transmitter side		At the receiver side	
Ne	4E 65	Ne	4E 65
tw	74 77	tw	74 77
or	6F 72	or	6F 72
k-	6B 00	k-	6B 00
addition	9D 4E	addition	9D 4E
Find the 16th complement as follows:		Add the checksum character	62 B2
Write the max. number in the base 16 + 1 = F F F F + 1	15 15 15 16	Final result	00 00
Write the calculated addition	9D4E	NO ERROR in the received message	
Subtract them to find checksum character	62B2		

Table 5–12: Checksum character generation in the case of no error

If the third bit of the character "w" changes from "1" to "0," then the receiver will get the word "Netsork" not "Network," which is obviously an error in the message. In this case, the receiver will receive the ASCII code of 73 instead of 77 for the character "w." Table 5–13 shows error detection using the checksum method.

At the transmitter side		At the receiver side	
Ne	4E 65	Ne	4E 65
tw	74 73	tw	74 77
or	6F 72	or	6F 72
k-	6B 00	k-	6B 00
addition	9D 4A	addition	9D 4E
Find the 16th complement as follows:		Add the checksum character	62 B6
Write the max. number in the base 16 + 1 = F F F F + 1	15 15 15 16	Final result	00 04
Write the calculated addition	9D4A	AN ERROR in the received message	
Subtraction	62B6		

Table 5–13: Checksum character generation in the case of an error

Cyclic-Redundancy-Check

The checksum redundancy error check is a reliable method of error detection for a single-bit error but not for burst errors. A burst error is an alteration of two or more consecutive bits in a frame of information. To resolve this problem, the cyclic-redundancy-check (CRC) method of error detection was developed. The CRC method is capable of detecting both single-bit errors and burst errors. For example, CRC-16 can detect all burst errors of 16 bits or less.

The computation of the CRC code is based on the polynomial division operation that employs module-2, where the remainder is determined by an XOR operation instead of the standard arithmetic division. A data bit stream is represented mathematically as a polynomial. The exponent of each term in a polynomial indicates the position of logic 1 in the data bit stream. For example, the X^6 term in a polynomial means the value of the 6th bit of the data bit stream is logic 1.

The CRC code is determined by division of the message polynomial M(x) by a specific polynomial which is called the *generator polynomial G(x)*. The reminder of the division is the CRC code and the quotient will be disregarded.

If there are *m* bits of message in *n* bits of transmission data in a cyclic block of codes (where $m < n$), then the length of the CRC code is equal to $n - m$ bits. The length or number of bits in a CRC code can also be found from the maximum power of the generator polynomial.

At the receiver's side, the message with the appended CRC code will be divided by the generator polynomial. A zero remainder is an indication of no error status.

There are a variety of CRC checking errors that are used in LAN and WAN protocols and also in error detection in compact disks and DVDs.

Table 5–14 shows the most common CRC generation polynomials and Figure 5–10 shows a block diagram of the CRC-16 circuit where each block is a D flip-flop.

CRC type	Generator polynomial
CRC-4	$X^4 + X + 1$
CRC-8	$X^8 + X^2 + X + 1$
CRC-16	$X^{16} + X^{12} + X^5 + 1$
CRC-32	$X^{32} + X^{26} + X^{23} + X^{22} + X^{16} + X^{12} + X^{11} + X^{10} + X^8 + X^7 + X^5 + X^4 + X^2 + X + 1$

© CENGAGE LEARNING 2012.

Table 5–14: Standard CRC generator polynomial

Example 9: Convert the message 10110011 to its corresponding polynomial.

Solution: Since bits zero, one, four, five, and seven are represented by logic 1, the corresponding polynomial is:

$$M(x) = X^7 + X^5 + X^4 + X^1 + X^0 = X^7 + X^5 + X^4 + X + 1$$

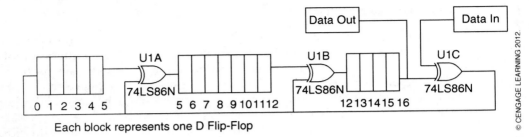

Each block represents one D Flip-Flop

Figure 5–10: CRC-16 block diagram

Example 10: Do the following operations if the message in Example 8 is set to transmission:

1. Determine the CRC code for the message in Example 8. Use the CRC-4 generator polynomial. The polynomial of CRC-4 is $X^4 + X + 1$.
2. Show there is no error if there is no bit change in the message at the receiver's side.
3. Show there is an error if a single bit (bit 3) has changed during transmission.
4. Show there is an error if two consecutive bits (bits 2 and 3) have changed during transmission. The number of consecutive bits k that have changed in the message is less than the maximum power of the generator polynomial n ($k = 2 < n = 4$).
5. Show there is an error if three consecutive bits (bits 2, 3, and 4) have changed during transmission. The number of consecutive bits k that have changed in the message is less than the maximum power of the generator polynomial n ($k = 3 < n = 4$).
6. Show there is an error if four consecutive bits (bits 2, 3, 4, and 5) have changed during transmission. The number of consecutive bits k that have changed in the message is the same as the maximum power of the generator polynomial n [$(k = 2) = (n = 4)$].
7. Show there is an error if five consecutive bits (bits 3, 4, 5, 6, and 7) have changed during transmission. The number of consecutive bits k that have changed in the message is greater than the number of bits in the generator polynomial n by one unit [$(k = 5) = (n = 4) + 1$].
8. Show there is an error if six consecutive bits (bits 1, 2, 3, 4, 5, and 6) have changed during transmission. The number of consecutive bits k that have changed in the message is greater than the number of bits in the generator polynomial $n + 1$ [$(k = 6) > (n = 5)$].

Solution: a. To find the CRC code, the bit stream in the message has to shift four times (the same as the maximum power of the generator polynomial) and then be divided by the bit stream of the generator polynomial:

Message polynomial = 10110011

Message polynomial after four shifts to the left = 101100110000

Generator polynomial = 10010

Mod-2 division of the shifted message polynomial by the generator polynomial is done by the XOR ($\otimes$) operation of these two polynomials.

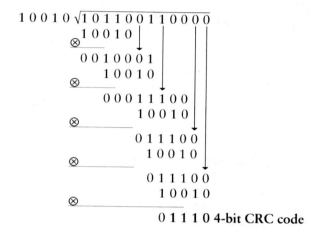

```
1 0 0 1 0 √1 0 1 1 0 0 1 1 0 0 0 0
         1 0 0 1 0
      ⊗ ‾‾‾‾‾‾‾‾‾
         0 0 1 0 0 0 1
             1 0 0 1 0
          ⊗ ‾‾‾‾‾‾‾‾‾
             0 0 0 1 1 1 0 0
                 1 0 0 1 0
              ⊗ ‾‾‾‾‾‾‾‾‾
                 0 1 1 1 0 0
                 1 0 0 1 0
              ⊗ ‾‾‾‾‾‾‾‾‾
                 0 1 1 1 0 0
                 1 0 0 1 0
              ⊗ ‾‾‾‾‾‾‾‾‾
                 0 1 1 1 0  4-bit CRC code
```

b. To find out if there is an error during transmission of the message, the CRC code must be appended to the end of the message and the same XOR operation above has to take place.

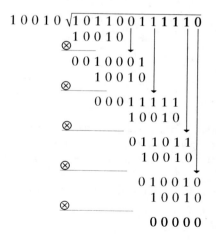

```
1 0 0 1 0 √1 0 1 1 0 0 1 1 1 1 1 0
         1 0 0 1 0
      ⊗ ‾‾‾‾‾‾‾‾‾
         0 0 1 0 0 0 1
             1 0 0 1 0
          ⊗ ‾‾‾‾‾‾‾‾‾
             0 0 0 1 1 1 1 1
                 1 0 0 1 0
              ⊗ ‾‾‾‾‾‾‾‾‾
                 0 1 1 0 1 1
                 1 0 0 1 0
              ⊗ ‾‾‾‾‾‾‾‾‾
                 0 1 0 0 1 0
                 1 0 0 1 0
              ⊗ ‾‾‾‾‾‾‾‾‾
                 0 0 0 0 0
```

The remainder is zero and therefore, no errors have occurred during the transmission.

c. Change bit 3 of the message bit stream and redo the above XOR operation.

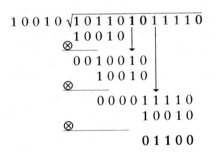

The remainder is not zero; therefore, there was a transmission error. This is a single-bit error detection case.

d. Change bits 2 and 3 of the message bit stream and redo the above XOR operation.

```
10010 √1011010111110
       10010 ↓      |
      ⊗_____  |
       0010010      |
        10010       |
       ⊗_____       ↓
        000011110
          10010
        ⊗_____
          01100
```

The remainder is not zero; therefore, there was a transmission error. This is a 2-bit burst error (two consecutive bit errors) detection case where the number of bits in the burst errors is less than the maximum power of the generator polynomial.

e. Change bits 2, 3, and 4 of the message bit stream and redo the above XOR operation.

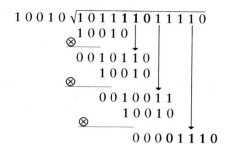

The remainder is not zero; therefore, there was a transmission error. This is a 3-bit burst error (three consecutive bit errors) detection case. All 3-bit burst errors can be detected since the number of bits in the burst error is less than the maximum power of the generator polynomial.

    f. Change bits 2, 3, 4, and 5 of the message bit stream and redo the above XOR operation.

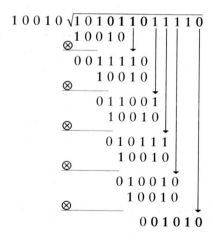

The remainder is not zero; therefore, there was a transmission error. This is a 4-bit burst error (four consecutive bit errors) detection case. All 4-bit burst errors can be detected since the number of bits in the burst error is equal to the maximum power of the generator polynomial.

    g. Change bits 3, 4, 5, 6, and 7 of the message bit stream and redo the above XOR operation.

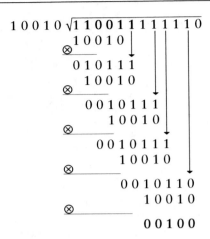

The remainder is not zero; therefore, there was a transmission error. This is a 5-bit burst error (five consecutive bit errors) detection case. Not all 5-bit burst errors can be detected because the number of bits in the burst error is greater than the maximum power of the generator polynomial by one unit ($k = n + 1$). The probability of not detecting a 5-bit burst error is $(1/2)^{n-1} \times 100 = (1/2)^{4-1} \times 100 = 12.5\%$. This means 8 out of 100 5-bit burst errors may not be detected by this generator polynomial.

h. Change bits 1, 2, 3, 4, 5, and 6 of the message bit stream and redo the above XOR operation.

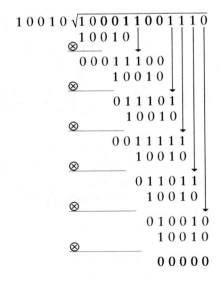

The remainder is zero, which means there is no error in the received message. This is obviously a false result and shows that the CRC method may not be able to detect all burst errors. If the number of bits in the burst error is greater than the maximum power of the generator polynomial plus one ($k > n + 1$), the probability of not detecting the burst error is equal to $(1/2)^n \times 100 = (1/2)^4 \times 100 = 6.25$ %. This means 16 out of 100 6-bit burst errors may not be detected by this generator polynomial.

The number of undetected burst errors decreases as the maximum power of the generator polynomial increases. Therefore, CRC-32 has much greater detection power than CRC-16.

Hamming Code

Hamming code, which was developed at Bell Laboratories by Richard W. Hamming, is an error detection and correction method that can detect two errors but correct only a single-bit error. Hamming code is capable of detecting a burst error but the process would be more complex and time-consuming. In this method, the error check bits, which are called Hamming bits, are randomly inserted into the data frame and then the positions of logic 1 are recognized and the XOR operation is performed on them. To reduce the chance of error in the Hamming bits, it is recommended to not place them all together. Hamming code cannot detect the errors if they occur in the Hamming bits. Insertion of the Hamming bits in the data frame at the receiver's side must be identical with the bits from the transmission side. The number of Hamming bits depends on the number of bits in the data frame and is calculated from the following equation.

$$2^n \geq m + n + 1 \tag{5.4}$$

where m is the number of bits in the data frame and n is the number of Hamming bits. The following example shows Hamming bits and the processes of Hamming code determination and error checking.

Example 11: Determine the number of Hamming bits, their logic values, and Hamming code for a 10-bit data frame of 1101001110. Assume a single-bit error has occurred during transmission at bit-6 and show how Hamming code will detect and correct it.

Solution: Use Equation (5.4) to find the number of Hamming bits.

$$2^n \geq m + n + 1$$
$$2^4 \geq 10 + 4 + 1$$
$$16 \geq 15$$

There are four Hamming bits: H_4, H_3, H_2, and H_1.

Randomly insert the Hamming bits into the data frame. Let's insert them between bit positions 2 and 3, 5 and 6, 7 and 8, and 9 and 10, as shown in Table 5–15.

Bit position	14	13	12	11	10	9	8	7	6	5	4	3	2	1
Hamming code	1	H_4	1	0	H_3	1	0	H_2	0	1	1	H_1	1	0

Table 5–15: Hamming bits and their bit positions © CENGAGE LEARNING 2012.

Table 5–15 shows there is logic 1 in bit positions 2, 4, 5, 9, 12, and 14. Convert these numbers to equal length binary numbers by performing the XOR operation on them in order to find the logic bits of the Hamming bits, and then show all these processes in Table 5–16.

Bit position	Binary equivalent number
2	0 0 1 0
4	0 1 0 0
5	0 1 0 1
9	1 0 0 1
12	1 1 0 0
14	1 1 1 0
	Use the XOR operation on the logic bits of each column to find the Hamming bits
Hamming bits	1 0 0 0

Table 5–16: The process of finding Hamming bits

The Hamming code (H-code) is: 1 1 1 0 0 1 0 0 0 1 1 0 1 0

The receiver will find its own H-code and perform an XOR operation with the transmitter's H-code; there is no error if the two H-codes are identical and the result is zero.

If bit-6 has changed from zero to one, then the binary equivalent of number 6 must be included in Table 5–17 and the XOR operation carried out with all binary numbers including the H-bits.

Bit position	Binary equivalent number
2	0 0 1 0
4	0 1 0 0
5	0 1 0 1
9	1 0 0 1
12	1 1 0 0
14	1 1 1 0
Hamming bits	1 0 0 0
6	0 1 1 0
The result of the XOR operation	0 1 1 0

Table 5–17: Error in bit-6

This result shows the bit in position 6 has changed during transmission and must be corrected.

5.5 ENCRYPTION AND DATA COMPRESSION

Encryption

Error checking methods are intended to protect data from alteration due to noise but do not address security of data that may contain sensitive or private information. Encryption of data before transmission and de-encryption at the receiver are used to transmit information securely. Encryption is the division of data into series of fixed blocks, which are called *ciphertext,* using a specific algorithm that prevents an unauthorized person from gaining access to the transmitted information. De-encryption at the receiver reverses the encryption process and recovers the information in its original form. Dividing the information into blocks of ciphers is done by using mathematical permutations on expansion, compression, or shuffling transformations. Scrambling of signals or substitution of numbers by letters are two simple methods of encryption.

To make sure the information will be received securely, an algorithm is used to develop an encryption key. The encryption key is appended to the information at the transmission side and a predefined decryption key at the receiver will reverse the encryption algorithm. The encryption key is either public or private.

The Data Encryption Standard (DES), which was developed by the National Bureau of Standards (NBS), divides the data bit stream into blocks of 64 bits that are enciphered separately. The encryption key is a 56-bit key to maintain the secure transmission of information. The 56-bit encryption key functions like a combinational key with 2^{52} or 7.2×10^{16} possible combinations. It is almost impossible to break the combination code of an encryption key.

Data Compression

Data compression is used to adjust the number of bits of transmitted characters in order to reduce the amount of data in the frame, consequently using the transmission bandwidth efficiently. Data compression is mostly used in transmission of alphanumeric characters. In plain text, some characters, such as the letter "a" or the letter "e," are used very frequently while other characters, such as the letter "x" or the letter "z," are used much less frequently. Data compression is a process that encodes the most frequently used letters in fewer bits and the less frequently used letters as higher bits. This compression technique is called the *Huffman coding method.*

SUMMARY

Several parameters must be thoroughly considered in designing a data communication system. A well-designed system is one that comes with reliability, recovery, security, and consistency. The performance quality of a data communication system as a whole also depends directly on its transmission lines and its surrounding area. Echo, crosstalk, pitch and frequency shifting, jitter, dropout packets, impulse noise, delay, bandwidth, bandwidth-delay product, and throughput are among the important factors that affect the performance of the transmission lines. Specific attention is focused on bandwidth efficiency and delay, because the ability to meet growing demand for fast access to considerable amounts of information at any time depends directly on these two factors more than any others.

One of the most efficient methods of transmission in a communication system is multiplexing several signals to send them over a single high data rate link. Frequency-division-multiplexing, time-division-multiplexing, and wavelength-division-multiplexing are three main types of multiplexing of several signals in a communication system. Among these three types, time-division-multiplexing is used most commonly in data communication systems. Multiplexing lower level digital signals to higher level digital signals is done by using TDM. Statistical time-division-multiplexing is an improved version of TDM for more efficient use of the single channel.

Several methods of error detection and correction have been developed to deliver information at the receiver side without error. The simplest method is using a parity generator and parity checker. This method can detect only a single-bit error, which is not an efficient way of detecting errors in a data frame that contains multiple bits. Longitudinal and vertical redundancy check is a two-dimensional method of error detection and correction. This method also has some limitation. Checksum is also a simple method of error detection that is based on summation of the numerical value of characters to generate a checksum character which will be appended to the transmission data for data detection. Checksum is used in XMODEM and IP. Cyclic-redundancy-check (CRC) is a reliable method of error detection for a single-bit error but not for burst errors. There are different variations on the CRC method, which is used mostly in the United States. CRC can detect up to two errors but can correct only a single-bit error. Another method of error detection that is commonly used is the Hamming code. Hamming code is capable of detecting a burst error but its process would be more complex and time consuming .

Encryption and data compression are techniques that are used in data communication for security of data transmission and efficient use of the communication bandwidth.

Review Questions

Questions

1. Briefly define *jitter*.

2. Briefly define *latency*.

3. Briefly define *crosstalk* and its impact on voice communication.

4. What is impulse noise?

5. Briefly define *bandwidth efficiency* and why it is an important parameter in data communication?

6. What is the main difference between data rate and throughput?

7. Why do we need to multiplex several signals?

8. How does FDM combine multiple signals into one?

9. Describe the hierarchy of FDM schemes.

10. How does TDM work?

11. Which multiplexing technique is used most commonly in data communication?

12. Describe the hierarchy of TDM schemes.

13. Define the function of F-bits in the DS-1 frame.

14. Define the function of C-bits in the DS-1 frame.

15. Define *T-1 lines*.

16. What is the main difference between synchronized TDM and statistical TDM?

17. Why do we need a buffer for statistical TDM?

18. Assigning time slots process is required in what type of TDM?

19. What is bit padding, why do we need it, and how does it work?

20. What is the purpose of the secondary multiplexer?

21. What is the purpose of the secondary demultiplexer?

22. How does a demultiplexer work?

23. How does a WDM combine multiple signals into one?

Continues on next page

24. Describe the similarities and differences between FDM and WDM.

25. What is the purpose of a guard band?

26. What is the difference between a parity checker and a parity generator?

27. What transmission method uses a parity bit?

28. How many bits can be detected by the LRC–VRC method of error detection?

29. Explain how checksum works.

30. Define *forward error correction*.

31. What are the most common types of CRC method?

32. What is the generator polynomial of CRC-4?

33. What is the application of Hamming code?

34. What is the purpose of using data comparison?

35. How does encryption work?

Problems

1. Find the total receiving time when 6.8 Mb of information is transmitted by a digital communication system with speed of 1.2 Gbps.

2. Repeat Problem 1 for a digital communication system with a speed of 64 Mbps. Find the time difference between these two digital communication systems.

3. Calculate the propagation times in a plastic optical fiber and a copper cable that are used for cross-country (3000 miles) communication.

4. Find throughput in a LAN network that is capable of transferring 20,000 packets of information in 30 seconds. Assume each packet of information contains 1200 bytes.

5. Find the bandwidth efficiency of a 128-K modem that sends a modulated signal over a telephone line.

6. Show how four computer terminals with data rates of 1.2, 1.2, 1.0, and 1.2 Mbps, respectively, can send their information through a TDM system. Draw your suggested system.

7. Show how four computer terminals with data rates of 28, 56, 56, and 28 Mbps, respectively, can send their information through a TDM system. Draw your suggested system.

8. Calculate the bit rate for DS-0.

9. Use a multiplexer to show how a DS-2 signal is generated by DS-0 and DS-1 signals.

10. Show the bit rate for a 1.544-Mbps T-1 line.

11. Find the odd and even parity bits for transmission of the letter H and show the transmitted data bit frame in the asynchronous serial transmission method.

12. The word "Link" is set to be transmitted. Determine the following:

   a) The LRC, VRC, and BCC bits using odd parity for both LRC and VRC.
   b) How the LRC and VRC method detects a single-bit error. Assume bit four (X2) of the character "k" has been altered during transmission.
   c) How the LRC and VRC method detects double-bit errors. Assume bit four (X3) of the character "i" and character "n" have been altered during transmission.
   d) How the LRC and VRC method detects a single-bit error. Assume bit four (X2) of the character "n" and bit five (X3) of the character "k" has been altered during transmission.
   e) How the LRC and VRC method detects a 4-bit error. Assume bit four (X4) and bit five (X5) of the characters "L" and "k" have been altered during transmission.

13. Determine the checksum character of the word "Link."

14. Find the checksum at the receiver if there is no error in receiving the message of Problem 11.

15. Find the checksum at the receiver if the LSB of the character "n" in Problem 11 is changed from "0" to "1."

16. Convert the data message of "11000110" to its corresponding polynomial.

17. Do the following operation if the data message of Problem 11 is transmitted over a digital communication system:

   a) Determine the CRC code for the data message in Example 9. Use the CRC-8 generator polynomial. The polynomial of the CRC-8 is: $X^8 + X^2 + X + 1$.
   b) Show there is no error if there is no bit change in the message at the receiver's side.

Continues on next page

c) Show there is an error if a single bit (bit 4) has changed during transmission.

d) Show there is an error if two consecutive bits (bits 3 and 4) have changed during transmission.

e) Show there is an error if three consecutive bits (bits 3, 4, and 5) have changed during transmission.

f) Show there is an error if four consecutive bits (bits 2, 3, 4, and 5) have changed during transmission.

g) Show there is an error if five consecutive bits (bits 1, 2, 3, 4, and 7) have changed during transmission.

h) Show there is an error if six consecutive bits (bits 1, 2, 3, 4, 5, and 6) have changed during transmission.

18. Determine the Hamming bit for the ASCII character "D."

19. How many Hamming bits are required for a single Unicode character?

20. Determine the number of Hamming bits, their logic values, and Hamming code for a 12-bit data frame of 111001011001. Assume a single-bit error has occurred during transmission at bit 8 and show how Hamming code will detect and correct it.

Experiment 5-1

Parity Generator and Checker

Introduction: Briefly write about parity bit, parity generator, parity checker, and the single-bit error detection method. Explain the function of each and how they are developed?

Parts and Equipment

- 74LS86 XOR gate (4)
- 74LS74 D Flip-Flop
- LED any color (2)
- SPDT switch (7)
- Power supply (5 V-DC)

1. Construct the parity generator and checker shown in Figure 5–11.

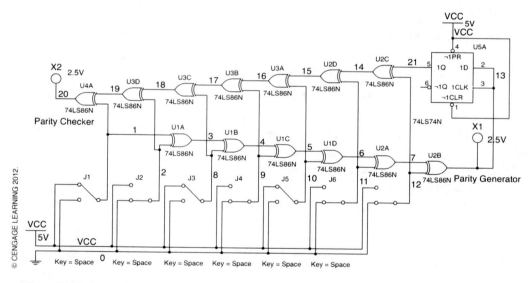

Figure 5–11: Parity generator and checker

2. Set the input switches on "1010100" to represent your selected data (the ASCII code for the letter T). Remember that logic 1 represents the ON state and logic 0 represents the OFF state.

3. Carry out an XOR operation on the seven bits of letter "T" to find the state of the parity generator bit.

Continues on next page

4. Determine the parity checker bit.

5. Turn ON the power supply.

6. Check the state of the parity generator bit and compare it with your calculated value in Step 3. Are they the same?

7. Check the state of the parity checker bit and compare it with your calculated value in Step 4. Are they the same?

8. Change the state of switch J2 from OFF to ON and observe the state of the LEDs at the parity generator and parity checker. Explain why they are ON or OFF.

9. Repeat Step 8 by changing the state of switches J2 and J5 and explain your finding.

Questions

1. How many error bits can be detected by the circuit shown in Figure 5–11?

2. Is the circuit shown in Figure 5–11 capable of correcting the detected error bit? Explain how.

3. What is the purpose of the D flip-flop in the circuit?

4. Can you use the circuit to detect an error bit for other codes such as Unicode that has 16 bits rather than the 7 bits of the ASCII code? Explain how.

Conclusion

Time-Division-Multiplexing (TDM)

Introduction: Time-division-multiplexing is a method of transmitting multiple digital signals over a single channel. TDM is used extensively in fiber optic communication.

Parts and Equipment

- 555 Timer (1)
- 4-bit binary counter, 74LS93N (1)
- Differential 4-channel multiplexer, MPC509AP (1)
- 1-kΩ resistor (4)
- 4.7-kΩ and 100-kΩ (one of each)
- 1-MΩ potentiometer
- 0.001-µF and 0.1-µF capacitors (one of each)
- 5-V DC power supply
- 4-channel oscilloscope

1. Construct the TDM circuit shown in Figure 5–12.

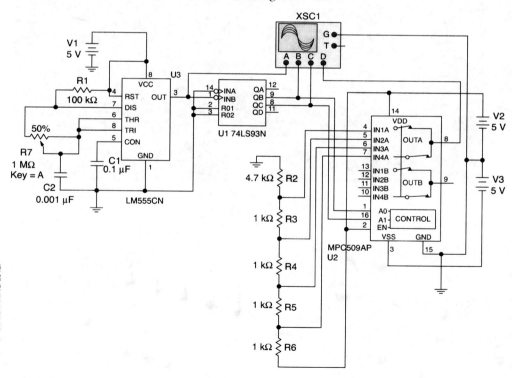

Figure 5–12

Continues on next page

2. Adjust the frequency of the 555 timer by setting the potentiometer at 50% of its capacity. Channel 1 of the 4-channel oscilloscope will show the waveform of the 555 timer (the first signal), channels 2 and 3 will show the second and third signals, and channel 4 will show the fourth signal of the modulated signals. The second and third signals are the output signals of the 4-bit binary counter and the fourth signal is the output signal of the 4-channel multiplexer. Figure 5–13 shows the first three signals. If you are building the circuit and your laboratory is not equipped with a 4-channel oscilloscope, you will need two oscilloscopes. If you want to simulate only the TDM circuit, depending on the version of the Multisim software, you will get all or some of the modulated signal.

3. Try to replace the 4-channel multiplexer (MPC509AP) with a Dual 3-channel analog multiplexer (4052) and write your findings in the conclusion.

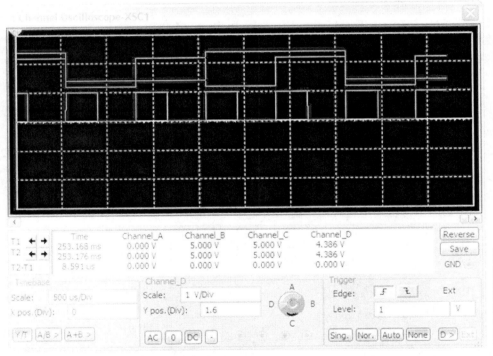

© CENGAGE LEARNING 2012.

Figure 5–13

Questions

1. What is the function of the 4-bit binary counter?

2. What happens if we use the first and the last bits of the 4-bit binary counter instead of the second and third bits?

3. Set the potentiometer at 10% of its capacity and observe the output signals. Write your observation in the Conclusion.

4. What is the function of the differential 4-channel multiplexer?

Conclusion

A Design Experiment: Single-Bit Error Detection and Correction Using the Hamming Code Method

Introduction: Briefly explain the Hamming code (H-code), H-code generation, error detection and correction.

1. Consider an 8-bit data bit of "11000100" as the information that needs to be transmitted.

2. Find out the number of H-code bits (n) from Equation (5.4):

$$2^n \geq m + n + 1$$

3. If the value of n is 4, then insert the H-code $(H_4 H_3 H_2 H_1)$ into the information bits in the positions 1, 2, 4, and 8, respectively.

$$(H_4)\ (H_3)\ (1)\ (H_2)\ (1)\ (0)\ (0)\ (H_1)\ (0)\ (1)\ (0)\ (0)$$

4. Determine the H-bits as follows:

    H_1 = XOR of bits: 3, 5, 7, 9, and 11

    H_2 = XOR of bits: 3, 6, 7, 10, and 11

    H_3 = XOR of bits: 5, 6, 7, and 12

    H_4 = XOR of bits: 9, 10, 11, and 12

5. Use XOR gates to design the H-code $(H_4 H_3 H_2 H_1)$.

6. Calculate the CHECKED bits (C-code) as follows:

    C_1 = XOR of bits: 1, 3, 5, 7, 9, and 11

    C_2 = XOR of bits: 2, 3, 6, 7, 10, and 11

    C_3 = XOR of bits: 4, 5, 6, 7, and 12

    C_4 = XOR of bits: 8, 9, 10, 11, and 12

7. Design a logic circuit to determine the CHECKED bits (you may use the H-code circuit to simplify your C-code circuit). You have to use only five 74LS86 ICs.

8. Simulate your H-code and C-code circuits. If all four bits of the C-code are 0, then there is no error in the transmission of the message.

9. Change bit-6 from logic 0 to logic 1 (low to high) and find the new H-code and C-code. Is there any error? Simulate the H-code and C-code circuits to verify your finding.

Questions

1. Carry out the XOR operation on all 12 bits of the data word plus the H-code. What is the result? What is the significance of the result?

2. What are the advantages and disadvantages of Hamming code over other error detection and correction methods?

3. How do you insert the H-code in the data bits?

4. Is the Hamming method capable of detecting an error burst? Explain how.

5. Name the other error detection methods and their advantages and disadvantages over the Hamming code method.

Conclusion

6

Network Standards

Objectives

After completing this chapter, students should be able to:

Describe the different network topologies.

Discuss network categories.

Describe the Open System Interconnection (OSI) model.

Discuss the IEEE standards.

Communication between nodes, terminals, or computers is possible by establishing a connecting system along with a set of rules that control and monitor the performance, reliability, flow, and error detection and correction of data packet exchanges. The connection system is called a *network*. Computers connect to the network via an attached interfacing device, called a *Network Interface Card* (NIC). A computer or a network node (in this chapter node will be used to represent any electronic device, including a computer, that is connected to the network) can also connect to another computer or network node directly via serial devices in a Point-to-Point Protocol (PPP). The set of rules that are needed to establish high-performance, error free, and reliable communication between computers or networks is called a *protocol*. The methods of connecting (wiring) several nodes together is called a *topology*.

To understand how nodes in a network or how entire networks can communicate with each other, we need to know the methods that they use to communicate, the methods of connection, and the rules that they have to follow.

6.1 PHYSICAL CONNECTION

Two or more nodes can be physically connected to each other using one of two different methods: point-to-point or multipoint connection.

Point-to-Point Connection

Point-to-point connection is a physical connection method in which a dedicated link connects two nodes directly via their Network Interface Cards. An RS232 interface cable or any other similar cable connects two nodes if they are located close to each other. Otherwise, data should be transferred through a modem. Point-to-point connection is also called a "store-and-forward" or "packet-switched" network, since a switch is able to connect several nodes in a point-to-point fashion through a technique called microsegmentation. In the point-to-point physical connection, each node is responsible for its own data formatting and connection. Figure 6–1 shows a typical point-to-point network connection.

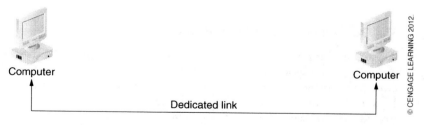

Computer   Computer

Dedicated link

© CENGAGE LEARNING 2012.

Figure 6–1: Point-to-point physical connection

Multipoint Connection

Multipoint connection is a physical connection method in which one node (the primary station or mainframe) is in charge of several other nodes (the secondary stations). The primary station can request information from the secondary stations at any time and the secondary stations will respond. Therefore, the secondary stations can send information only if the primary station has requested it. Two secondary stations cannot send information at the same time, otherwise there would be a collision between the transmitted information. Multipoint connection is also called *multidrop connection* because each secondary node drops (hangs) from the main link. In this method, the secondary nodes are connected to the main link via a T-connector. Figure 6–2 shows a typical multipoint connection.

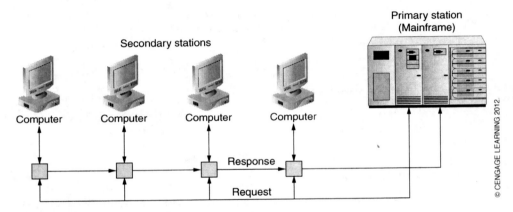

Figure 6–2: Multipoint connections

6.2 NETWORK TOPOLOGY

Topology is the physical layout of a network. Topology defines how the computers, nodes, or other network devices are arranged and wired together within a network system. The most common network topologies are: tree, bus, star, ring, mesh, and hybrid.

The Tree or Hierarchical Network Topology

This type of topology consists of a central station, which is the most powerful computer in the network (i.e., the mainframe, the root of the tree). All other nodes are connected to the central station in a hierarchical manner. In the hierarchical or tree layout, the central station (which is at the highest hierarchical level—level I) is connected to some secondary nodes, which are directly under its command. The secondary nodes are at the second highest hierarchical level (level II). Each of the secondary nodes, in turn, is connected to several other nodes that are under its command. These nodes are at the third highest hierarchical level (level III). These changes of command will continue until all nodes in the network are covered.

A higher-level node has access to the information of only those nodes that are directly under its command. Each lower-level node can access the information of other nodes only by requesting it from its higher-level node. For example, if a node at level III needs information from another node at level III but is under a different level II node, it needs to request the information from its own level II node and the level II node has to request it from the level I node. This network method is very inefficient because most of the nodes are at the very bottom of the hierarchy. Figure 6–3 shows a typical tree network topology with three hierarchical levels.

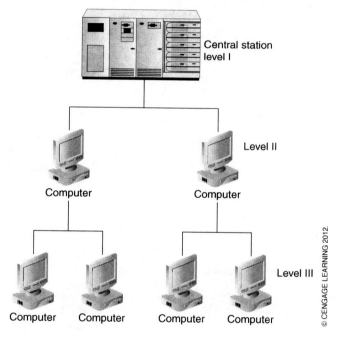

Figure 6–3: Tree network topology

The Bus Network Topology

This is a multiple-connection network where all nodes are connected to the network via a drop line and T-connector (tap) to a single cable which is also called the *backbone, bus,* or *trunk.* The backbone must be terminated at its end point by a terminator. The bus topology was (and still is in some specific applications) used in local area networks (LANs) because of its simplicity, easy implementation, requirement for less cabling, and cost-effectiveness. But advantages usually come with some disadvantages. The disadvantages of bus topology include: implementing limited nodes, since the signal in the network will be faded (high attenuation) as the length of the backbone gets longer. The entire network has to be closed down in order to add or remove any node in the network. Any break in the backbone will cut the communication to nodes that are not in the central station, and because the backbone is not terminated, noise will enter the entire network and disrupt all communication. Figure 6–4 shows the typical bus network topology.

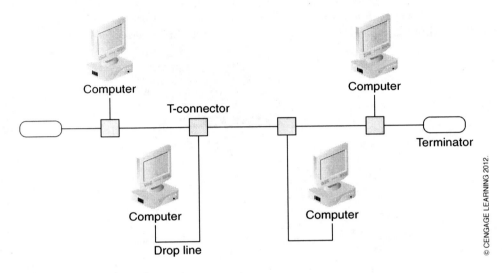

Figure 6–4: Bus network topology

The Star Network Topology

This type of network uses a hub as a central controller for the network. This means the hub is the heart of the network and all other nodes are connected to the hub via a point-to-point connection. This easy type of connection requires only one I/O for each node, which brings the cost of the network down. Because of the point-to-point connection between the nodes and hub, each node can only communicate with the hub. If any node needs information from another node in the network, it has to request it from the hub. Star topology is used in LANs and high-speed LANs.

Advantages of the star topology include ease of adding or removing a node without closing down the entire network or interrupting communication; if one node fails to operate, the other nodes will function without interruption; and it is easy to troubleshoot and isolate the fault in the network. The disadvantages include that if the hub fails then the entire network fails. This topology also requires more cabling than some others, such as the bus topology. Figure 6–5 shows a typical star network topology.

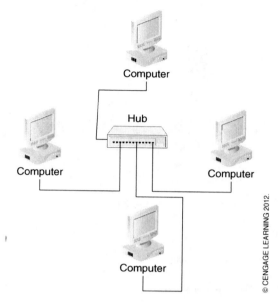

Figure 6–5: Star network topology

The Ring Network Topology

The ring network typology was introduced by IBM for its Token Ring LAN technology, which will be discussed in detail in Chapter 8. Each node in the ring topology is connected via a point-to-point connection to two nodes to create a ring-shaped network. Any node can be a source to transmit information by a token that circulates along the ring in only one direction. All nodes in the ring starting with the node next to the source in the direction of the token will examine the destination address in the token. If it is matched with its own address, it will copy the information in its buffer and release the token into the ring. If the destination address does not match with that particular node address, it will release the token into the ring without copying the information. This process will be repeated until the information reaches its destination.

Ring topology uses the multiple-station-access-unit (MAU) for token circulation in the ring. The advantages of ring topology are relatively easy installation, diagnosing faults, and troubleshooting. The disadvantages are expansion of the network and the interruption of the entire network if the ring breaks. Besides Token Ring technology, the Fiber Distributed Data Interface (FDDI) also uses ring topology. Figure 6–6 shows a typical ring network topology.

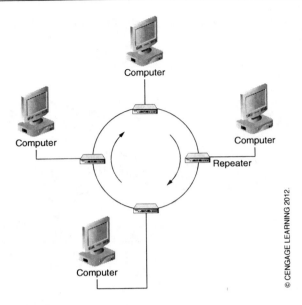

© CENGAGE LEARNING 2012.

Figure 6–6: Ring network topology

The Mesh Network Topology

This topology allows every node to be connected to all other nodes in the network. All connections are point-to-point and need dedicated links. Since each node has to be connected to all the other nodes, it requires the same number of I/O ports as the number of nodes in the network. Therefore, this topology requires a lot of cabling and devices (i.e., I/O ports), which makes it very expensive and difficult to install and troubleshoot. For this reason, mesh topology is rarely used in network systems. The advantages of the mesh topology are that if one node is down, all other nodes will function fully and, because there are many alternative routes to deliver the information, data will continue to be transmitted in the network. Equation (6.1) shows how to find out how many cables are needed to connect to implement a mesh topology for n number of nodes.

$$\text{Number of connecting cables} = \frac{n(n-1)}{2} \tag{6.1}$$

Example 1: How many cables are needed to implement a 10-node mesh network topology?

Solution: $\text{Number of cables} = \dfrac{10(10-1)}{2} = 45$

Example 2: How many nodes are arranged in a mesh network topology with 105 cables?

Solution: $105 = \dfrac{n(n-1)}{2}$

$$(105)(2) = n^2 - n$$

$$n^2 - n - 210 = 0$$

$$n = 15 \text{ nodes}$$

The Hybrid or Combination Network Topology

This is a mix of two or more other types of network topologies. Hybrid topology is usually used when a network layout already exists but needs to be expanded using a different type of layout. For example, suppose there is a bus network topology in our building with four nodes but we want to extend it with four more nodes that will be used in only one of the offices. The network engineer suggests that the star topology is more suitable for our needs and will cost less because all we need to do is connect a hub with four output ports to one of the original nodes and connect four new nodes to the hub. By using the hybrid topology in this situation, there would be no need to interrupt the network, remove any node, or extend the existing backbone. Figure 6–7 shows the hybrid network topology implementation for this example.

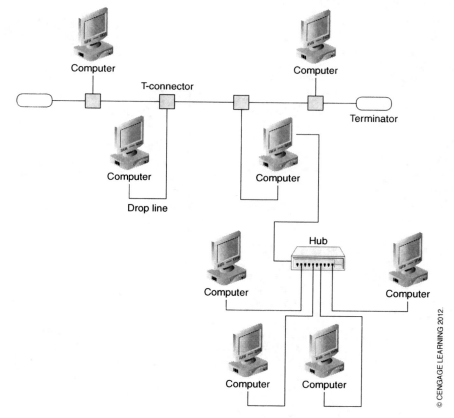

Figure 6–7: Hybrid network topology

6.3 NETWORK CATEGORIES

In a network system, a computer either provides services or can demand and use services. In the first case, the computer is called the server and in the second case, the computer is called the client. A network system is either a client-server or peer-to-peer network.

Client-Server Network

A client-server network is one in which one computer acts as a primary service provider and other computers are clients that use that service for their own specific applications. File, mail, and print servers are the most common servers in a LAN. Servers and clients are connected to each other by a LAN switch.

A file server accepts files from clients, stores them in its secure storage unit, creates a backup file, and allows clients to access the authorized file at any time

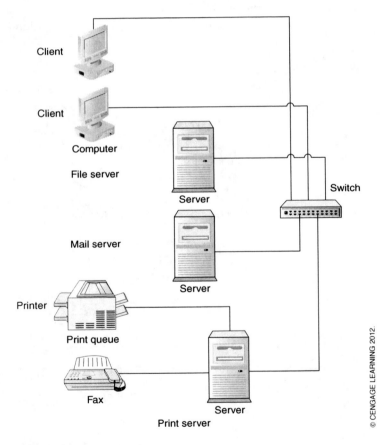

Client

Client

Computer

File server

Server

Mail server

Server

Switch

Printer

Print queue

Fax

Server

Print server

© CENGAGE LEARNING 2012.

Figure 6–8: A typical client-server network

without carrying the files in a personal storage unit such as a flash drive. To secure files from unauthorized clients, a user name is assigned to each individual client. In a client-server network, clients are able to share their files with others. One of the advantages of the file server is purchasing and storing a single copy of needed software instead of multiple copies.

A mail server does multiple tasks. It provides a mail address and user name for each client, distributes mail to the users, and communicates and exchanges mail with servers that may or may not have the same electronic mail programs. A mail server can notify the mail sender if the receiver is "out of office" or on "vacation." It is also able to provide a "request for response" or similar options for senders.

A print server provides access to all or select clients to access available printers, faxes, or copy machines. Figure 6–8 shows a typical client-server network.

Peer-to-Peer Network

A peer-to-peer network is one in which each computer in the network can both provide and use services. This means each computer can be a client and a server at the same time. Each computer has to decide which other computers have the right to access its files and which do not. Consequently security and administration become an issue in this type of network setting and each computer needs to do its own administration tasks and set its own policy on security issues. A peer-to-peer network functions well for connecting a few computers (typically around 10 computers) and all computers have to have their own network card. In this type of network, computers are connected to each other by a cable via a hub or switch. Figure 6–9 shows a typical peer-to-peer network setting.

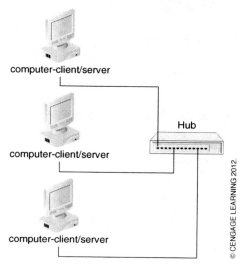

Figure 6–9: A typical peer-to-peer network

6.4 NETWORK PROTOCOLS

Network protocols are collections of rules and regulations that oversee all operations within a network so that two devices can exchange information by understanding and accepting the way each side forms and processes information and transmits it. There is no unique protocol that provides a well-rounded set of rules to ensure reliable communication between two computers or other communication devices. To have well-rounded communication, usually a group of protocols, commonly called a protocol family, is used in the network operation. The rules of a protocol may define the following characteristics of a network for proper and reliable communication:

1. Type of network topology
2. Type of signaling
3. Type of cabling
4. Speed of data exchange
5. Size of the information
6. How to access the information
7. How data should be formatted and presented (syntax)
8. The meaning or language of the data (semantics)
9. Validation of the information
10. Method of error detection and correction
11. Synchronization between transmitter and receiver

The performance of the above tasks is very complex and difficult. To ease and smooth the performance of a network protocol, it is divided into layers where each layer performs a specific task. Network protocols are developed by a variety of organizations but the two most common are:

1. Transmission Control Protocol/Internet Protocol (TCP/IP). TCP/IP is a four-layer protocol consisting of link, network, transport, and application layers. The TCP/IP protocol is discussed in detail in Chapter 11.
2. Open System Interconnection (OSI) Protocol. OSI is a seven-layer protocol that consists of physical, data link, network, transport, session, presentation, and application layers.

Open System Interconnection

OSI is a reference model that was developed by the International Organization for Standardization (ISO) about three decades ago in order to establish a set of standards that allow the variety of existing communication devices and applications to be able to exchange information in a network system. The seven layers of the OSI model are divided into two groups of upper and lower layers. The four lower layers are responsible for moving data, and they consist of both hardware and software. The top three layers are responsible for application issues implemented using software. The combination of the first three layers is known as the media layers and the combination of the last four layers is known as the host layers.

Data exchange between two computers usually starts at the application layer (the seventh layer) and passes through the layers from top to bottom to become a transmittable signal over a specified transmission channel. Data generated at the top layer will be divided in segments when they leave the host layers and segments of data will be changed to packets, frames, and bits in the media layers before transmission over a link that connects two computers. In addition to these processes, the OSI model will help the transmitted data to be defined in the right format, size, speed, error and flow control type, route, and all other characteristics in order to arrive at the right destination. Figure 6–10 shows how data will be processed for transmission and how it arrives at the right destination based on the OSI model.

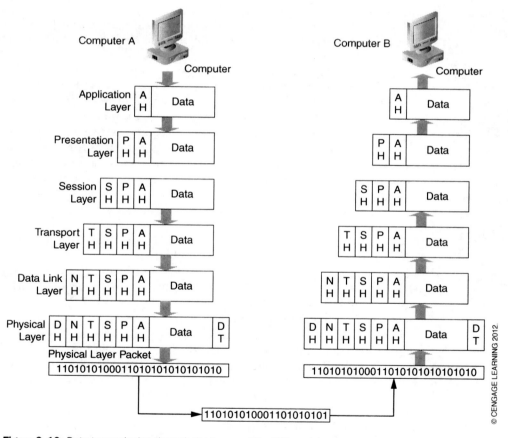

Figure 6–10: Data transmission through the layers of the OSI model

As shown in Figure 6–10, each layer adds its own header to the previous layer frame (which is also called the protocol data unit, PDU) and creates a new PDU. For example, the application layer adds it own header (application header, AH) to the data and creates the application PDU, which it passes to the presentation layer. The AH contains information that belongs to the application layer only; therefore, other layers do not open the AH. They only add their own header (which is called the *encapsulation process*) and pass it down to the next lower layer until it gets to the physical layer. The physical layer passes the data link PDU to the next communication device which understands the data link PDU and accepts it. The data link PDU of the sender will move up from layer to layer on the receiver side, starting from the physical layer. Each layer opens and reads its own appropriate header, removes it from its own PDU, and passes it to the next upper layer. In addition to the header, the data link layer adds another field to its layer frame that is called a *trailer* (data link trailer, DT). The PDU in the data link layer is called the LPDU or simply a frame.

The data link trailer (DT) usually contains the error-checking bits (in the form of the frame check sequence, FCS) to confirm the accuracy of the transmitted data.

The physical layer packet does not always move directly from one computer to another. It may go to an intermediate device that only facilitates the passage of the packet from one computer to another. The intermediate devices may operate at different layers of the OSI model; for example, routers operate at the network layer (layer-3). Therefore, when data enters a router, it passes through the physical and data link layers before entering the network layer for redirection by the router. Figure 6–11 shows the process of data transmission between two computers via a router. As shown in Figure 6–11, each layer at the transmitter side is connected to its counterpart at the receiver side via a virtual link, but the connections between the transmitter, intermediate device, and receiver are made by a real (physical) link.

Industrial data communication is not as complex as general data communication. The two lower layers and the top layer of the OSI model (the physical, data link, and application layers) are able to establish communication between industrial devices (instruments and controllers) without requirement of other layers. Fieldbus technologies like PROFIBUS and FOUNDATION™ Fieldbus comprise the physical, data link and application layers.

Fieldbus technology is a digital communication network standard to replace the 4-20 mA current loop analog standard and provides real-time distributed control in the instrumentation and control industry. A fieldbus technology network is a bidirectional, multipoint, serial communication that is able to accommodate daisy chain, ring, star, and tree network topologies. In the fieldbus technology, field devices such as sensors, transducers, and controllers are made into smart devices by installing low-cost computing power into them. These smart devices are able to diagnose,

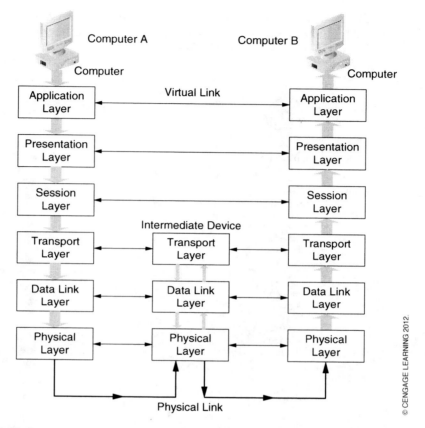

Figure 6–11: Data transmission via an intermediate device

control, and maintain their function and can communicate with other smart devices by connecting to them in a bidirectional fashion.

The following eight different communication protocols, called *types,* have been developed for the fieldbus technology:

- Type 1 FOUNDATION Fieldbus H1
- Type 2 ControlNet
- Type 3 PROFIBUS
- Type 4 P-NET
- Type 5 FOUNDATION Fieldbus HSE (High-Speed Ethernet)
- Type 6 SwiftNet (a protocol developed for Boeing, since withdrawn)
- Type 7 WorldFIP
- Type 8 INTERBUS

Figure 6–12 shows the modified OSI model that is used in the fieldbus and FOUNDATION Fieldbus network technologies. The functions and tasks of each layer in the OSI model are described below.

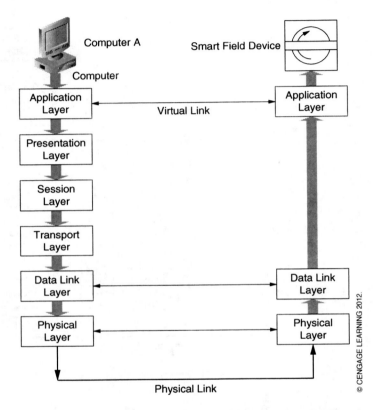

Figure 6–12: Data communication in the fieldbus and FOUNDATION Fieldbus network technologies

The Application Layer

The application layer defines the language and its rules that programs need to use to establish communication between client and server (using protocols such as the Association Control Service Element, ASCE). The application layer helps clients to access any file by logging on to the remote host, opening the file, reading or writing to it, transferring files, obtaining information, and executing a remote job. Simply put, the application layer provides a user interface via a network virtual terminal and supports network-based services such as access to file, file transfer, and mail services through its protocols. To provide the user interface and services, the application layer identifies communication partners, determines resource availability, and establishes cooperation between partners. For example, if a client wants to access an Internet site, the client writes the site address and clicks the "Enter" key. Meanwhile, the Internet application prepares a file with the address and passes it to the HTTP protocol that exists on the local server. The HTTP formats the application layer PDU and passes it to the lower layer. The application PDU will pass through each layer as described earlier to finally reach the destination. The destination will read the message, let the client open the site and access and retrieve the needed information.

The following are some of the TCP/IP protocols that are implemented in the physical layer: Simple Mail Transfer Protocol (SMTP), Real-Time Transfer Protocol (RTP), Domain Network System (DNS), Terminal Network (TELNET), and Hypertext Transfer Protocol (HTTP).

The Presentation Layer

The presentation layer presents transparent data to the application and session layers in order to establish exchangeable information between two systems that use different syntax and semantics. The presentation layer will translate and interpret data from the sender's specific format into the commonly used format. The main functions of the presentation layer are: to convert EBCDIC-coded or Unicode-coded text to ASCII-coded text, to convert encrypted information to decrypted information, and to convert compressed data to decompressed data that is used widely in audio and video in order to provide usable data to the destination. The presentation layer also controls security at the file level. The header at this layer contains the characteristics and length of the transmission.

The Session Layer

The session layer establishes and manages the connection. The main functions of the session layer are that it identifies the source and the destination and then establishes a connection between them; controls dialogues between them (initializes the start of the session, performs the login process and request for username and password, and manages the orderly closing of a session); places the synchronization point (checkpoint) to stop resending the part of the information that has arrived at the destination without an error; and facilitates the simplex, half-duplex, and full-duplex operations. The header at this layer contains the control information.

The Transport Layer

The transport layer provides a reliable end-to-end or process-to-process error-free message delivery via flow control, segmentation and desegmentation, and error control in a network. The transport layer accepts the session layer PDU (SPDU), segments it logically, and adds its own header to create several transport PDUs (TPDU) in order to match them with the network layer frame size. The desegmentation process will take place at the destination's transport layer.

Error checking at the transport layer occurs at the segment level. The transport layer monitors the segments, corrects them if they are out of order, and retransmits the corrected segments. This means the error correction is achieved by retransmission of the failed segments.

The most common protocols of the transport layer are Transport Protocols class 0 to class 4 (TP0, TP1, TP2, TP3, and TP4). TP4 is designed for a less reliable network. TP4 is capable of providing both connection- and connectionless-oriented packet delivery, segmentation and desegmentation, flow control, error recovery, multiplexing and demultiplexing over a single virtual circuit, and reliable transport service.

The most common TCP/IP protocols that are implemented in the transport layer are TCP and UDP. The User Datagram Protocol (UDP) provides a direct method of sending and receiving datagrams over an IP network.

The Network Layer

The network layer provides data delivery between source and destination across multiple networks. The network layer determines how data can be forwarded from the source to its destination by finding a path through a series of switching points (routers). In a sense, the network layer provides connection, disconnection, routing, and multiplexing. The transport layer accepts the SPDU and adds its own header that contains the logical addresses of the source and the destination to create the NPDU. The NPDU may be fragmented into smaller units if the networks in the communication route have different data unit sizes. The network layer provides both connection- and connectionless-oriented services.

The network layer flow control feature makes sure the destination is fast enough to store data into its buffer. It will stop transmission if the destination is not fast enough to catch up with the source's transmission speed.

The IP protocol is implemented in the network layer. AppleTalk, IPX, and SNA are among the routable protocols that are implemented in this layer.

The Data Link Layer

The data link layer is one of the most important layers in the OSI model. It was originally developed for point-to-point and point-to-multipoint communication, such as a telephone system. The data link layer accepts the NPDUs and divides them into frames, and adds the destination and source addresses, the error-checking field (frame check sequence, FCS field), and the data length field to the frame to create the data link PDU (LPDU), which is sent to the physical layer for transmission. At the destination side, the data link layer accepts bits from the physical layer, changes them into the frame and checks for any possible error at the bit level. If there is no error, the frame will be sent to the network layer.

The other functions of the data link layer are to: establish frame synchronization between the source and destination, flow control to make sure the destination adjusts its capacity in line with the incoming data rate to prevent overload, distinguish between the data frame and the control frame, and manage the communication between networks.

The High Level Data Link (HDLC) which is a bit-oriented protocol for managing information, and the Link Access Procedure Balanced (LAPB or X.25) protocol, a modified version of HDLC, are the most common data link protocols.

HDLC was developed by the ISO and is based on the IBM Synchronous Data Link Control (SDLC) protocol. HDLC is a widely used protocol for the data link layer and is a bit-oriented protocol that supports both half- and full-duplex communication, point-to-point or multipoint networks, and switched and nonswitched channels to establish synchronous data transmission between two nodes. The physical layer provides a clocking and synchronization method for the HDLC protocol. Figure 6–13 shows the HDLC protocol frame format

Flag	Address	Control	Information	FCS	Flag

Figure 6–13: HDLC frame format © CENGAGE LEARNING 2012.

The HDLC protocol defines three nonoperational modes: Normal Disconnected Mode (NDM), Asynchronous Disconnected Mode (ADM), and Initialized Mode (IM).

There are also three different types of stations that can be used in the HDLC protocol: primary, secondary, and combined. Primary stations can send commands, the secondary stations are only able to respond, and in the combined type, both primary and secondary stations can send and respond.

The flag or frame delimiter fields are 8-bit (01111110 = 7E in hexadecimal) fields to indicate the beginning and end of the frame. To prevent repetition of the flag field in the information field, the bit stuffing mechanism is used. Bit stuffing means inserting a bit 0 whenever a series of five 1 bits occur in the information field. The inserted bit is removed at the receiver side.

The address field is usually an 8-bit field. The first bit in the address field indicates the type of recipient station. If the first bit is logic 1, the recipient station is broadcast or multicast, and if it is logic 0, the recipient station is unicast.

The control field is either 8 or 16 bits and it contains the command, response, and sequence numbers to maintain data flow. The control field also defines the frame types. The three frame types in the control field are: information frame (I-frame), supervisory frame (S-frame), and unnumbered frame (U-frame). These three will be discussed in detail at the end of this chapter.

The data or information field is variable. The FCS field is either a 16-bit or 32-bit CRC method.

The Point-to-Point Protocol and the Serial Line Internet Protocol are TCP/IP protocols that are implemented in the data link layer. The IEEE standard has divided the data link layer (DLL) into two sublayers, the Logical Link Control (LLC) and Media Access Control (MAC) and developed the IEEE 802.x standard. The IEEE standard for the data link layer will be discussed in the next section.

The Physical Layer

The physical layer defines the procedure for transmitting a signal. Basically, the physical layer is divided into two main sublayers: a lower layer that is called Physical Media Dependent (PMD) and the upper level that is called Physical Media Independent (PMI). The PMD itself divides into several sublayers and the function of these sublayers is to determine the physical characteristics of the signal (frequency of operation, voltage level, optical level for optical communication, and signal duration or data rate), media (type of cable, i.e., coaxial, twisted pair, and optical fiber), connectors and interfaces (point-to-point or multipoint), direction of transmission (simplex, half-duplex, and full-duplex), and the network topology (star, ring, tree, mesh, bus, or hybrid). Bit presentation is the main function of the PMI sublayer. PMI accepts bits from the physical layer and converts them to the data frames that are acceptable for the receiver network.

The balanced serial interface standard (RS530 or EIA-530), EIA/TIA-232, and EIA/TIA-449 are the most common OSI protocols for the physical layer.

6.5 IEEE STANDARDS

It used to be difficult and sometimes even impossible to establish communication between the various network devices and systems in the market. This was especially due to the lack of standards among manufacturers that had designed, developed, and built these devices and systems. As a result, the Institute of Electrical and Electronic Engineers (IEEE) was chosen to develop a standard to enable all devices and systems to communicate together. In 1985, IEEE formed the IEEE 802 working group to develop such a standard in the OSI model framework. The IEEE 802 standards, or simply the IEEE standards, are the result of the IEEE 802 working group. Later, the IEEE standard was approved and adopted by the American National Standard Institute (ANSI) and the International Organization for Standardization (ISO).

The IEEE standards divide the data link layer of the OSI model into two sublayers: the lower sublayer, or Media Access Control (MAC), and the upper sublayer, or Logical Link Control (LLC). The MAC sublayer defines how a station or computer accesses the network. There is no unique access method for all types of networks.

The Ethernet employs the Carrier Sense Multiple Access with Collision Detection (CSMA/CD) method, and the Token Bus uses the token passing method. The LLC carries the information and makes connection possible between networks that function under different protocols. Figure 6–14 shows the relationship between the IEEE 802 LAN standards and the OSI model.

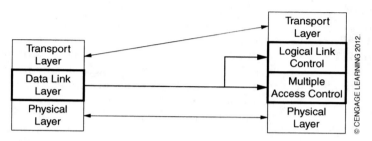

Figure 6–14: The OSI model and the IEEE 802 standards

In years following 1985, the IEEE 802 standard expanded to all types of network systems by specifying the physical and data link layers for each type of network. Table 6–1 shows the list of the IEEE standards for different types of network systems.

IEEE 802 standard	Description
802.1	Internet working (network management)
802.2	Logical Link Control (LLC)
802.3	Ethernet WG
802.4	Token Ring LAN WG
802.5	Token Ring LAN WG
802.6	Metropolitan Area Network (MAN)
802.7	Broadband Technical Advisory Group (TAG)
802.8	Fiber optics TAG
802.9	Integrated Service LAN WG
802.10	Interoperable LAN Security (SILS) WG
802.11	Wireless LAN WG
801.12	Demand priority WG
802.14	Cable modems
802.15	Wireless personal area network (PAN)
802.16	Broadband wireless access
802.17	Resilient Packet Ring (RPR) WG
802.18	Radio regulatory TAG

Table 6–1: IEEE standards

The Media Access Control Layer

The media access control layer (MAC) is the IEEE 802 standards designation for the lower sublayer of the data link layer in the OSI model. The MAC layer defines or regulates the methods that stations can use to connect to the network via a virtual point-to-point connection. The MAC layer reads the type of media (i.e., twisted pair or optical fiber) that the physical layer uses and responds to it accordingly. The access methods may vary from one network to another. For example, the access method for an Ethernet LAN network is the Carrier Sense Multiple Access with Collection Detection (CSMA/CD) or Token passing mechanism for the Token Ring LAN network.

The frame format of the MAC layer changes slightly from one MAC protocol to another. In general, the MAC layer frame format consists of the following fields:
1. The control field that monitors the proper operation of the MAC protocols.
2. The destination address field, which is a unique serial number that is assigned by the device manufacturer to identify the destination for the correct delivery of the information.
3. The source address field which is also a unique serial number that is assigned by the device manufacturer to identify and distinguish the source device among all other devices in the network.
4. The Logical Link Control Protocol Data Unit (LLC PDU).
5. The Frame Check Sequence that is calculated by the sender is used for error detection.

The first two fields are appended to the header and the last field is appended to the trailer of the MAC layer. The third field (LLC field) is placed between the header and trailer, which is also called the MAC payload. In a sense, the MAC layer does some framing work within the IEEE 802 standards.

The Logical Link Control Layer

The logical link control layer (LLC) is the IEEE 802 standard for the upper sublayer of the data link layer of the OSI model. Unlike the MAC layer protocol which varies from one network to another, the LLC layer protocol is a unique protocol for all IEEE 802 standards. Therefore, the LLC layer is independent of the MAC layer protocol and provides connection (either connection- or connectionless-oriented) service between networks that run under different protocols (i.e., TCP/IP, Windows XP, Novel IPX, or IBM Netbios). The other functions of the LLC layer are: frame formatting, source and destination address recognition (SSAP and DSAP), flow control, and error control. The header of the LLC layer includes DSAP, SSAP, and the control fields.

The Destination Service Access Point

The destination service access point (DSAP) field is an 8-bit field that identifies the service access point (SAP) of the destination's network protocol and sets up the connection accordingly. Each protocol has its own unique SAP. The first bit of the DSAP indicates the numbers of destinations, single or multiple destinations.

The source service access point (SSAP) field is also an 8-bit field that identifies the SAP of the source's network protocol and sets up the connection accordingly. The first bit of the SSAP indicates if the frame is a command or response type. Table 6–2 shows the selected SSAP and DSAP code assignments for different protocols.

SSAP and DSAP code (address) assignment	Protocol
06	ARPANET Internet Protocol (IP)
F0	IBM NetBIOS
E0	Novell NetWare
18	Texas Instruments
AA	SubNetwork Access Protocol (SNAP)
F5	IBM LAN Management (group)
4E	EIA RS-511 Manufacturing Message Service

Table 6–2: Selected SSAP and DSAP code assignments

The control field indicates the type of information field that can be either an information transfer frame format (I-frame), supervisory frame format (S-frame), or unnumbered frame format (U-frame).

The information frame (I-frame) utilizes the transformation of the information via a connection oriented mode. As shown in Figure 6–15, the frame format of the I-frame consists of four subfields.

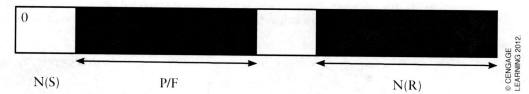

N(S)   P/F   N(R)

Figure 6–15: The frame format of the I-frame

The first subfield is a single bit (bit 0) that indicates the type of frame (I-frame). The next subfield that is shown by N(S) indicates how sequences of the frame have been sent. In regular I-frame format, the N(S) is a 3-bit subfield that contains eight sequences. The third subfield (P/F) is also a single-bit field. The P/F bit is one (logic 1) if the source (primary) is requesting its frame from the destination (secondary). The P/F bit is zero (logic 0) if the destination (secondary) is requesting its frame from the source (primary). The fourth subfield is shown by N(R) which indicates how many frames are expected to be received. The I-frame uses the piggybacking method for the received frame acknowledgment. The piggybacking method is used in bidirectional communication to acknowledge receipt of the transmitted frames without seeking the permission of the source (subscribers). In a sense, piggybacking is not an appropriate (or legal) access of the information.

The supervisory frame (S-frame) responds to the issue of inappropriate access to the information using the piggybacking method of acknowledgment. The S-frame is divided into four subframes. The first two bits are always 10 to indicate that it is an S-frame. The code subfield (a 2-bit subfield) defines the type of S-frame. There are four types of S-frames in terms of commands and responses, as shown in Table 6–3.

Code subfield	S-frame types	Acknowledge (ACK) types
00	Receive ready (RR) indicates that the frame or frames are received safely	Positive (ACK)
01	Reject (REJ) informs the sender before its time expires to resend the frame or frames	Negative (NACK)
10	Received not ready (RNR) informs the sender about receiving the frames but the sender must stop or wait because either too many frames are being received or there is a problem that needs to be fixed	Positive (ACK)
11	Selective reject (SREJ) is a negative acknowledgment that rejects a selected frame in order for it to be sent later.	Negative (NACK)

© CENGAGE LEARNING 2012.

Table 6–3: S-frame types

The fourth subfield is shown by N(R), which indicates the number of ACK and NACK, or how many frames have been received safely and how many have not. The S-frame does not have an information field. Figure 6–16 shows the control field for the S-frame.

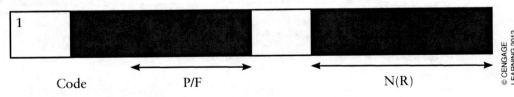

| Code | P/F | N(R) |

© CENGAGE LEARNING 2012.

Figure 6–16: The frame format of the S-frame

The unnumbered frame (U-frame) is concerned with the operation modes of the communication. The U-frame is also divided into four sublayers. The first two bits of the U-frame are 11, which indicate that this is a U-frame. Like the I-frame and S-frame, the U-frame also has a P/F subfield. There are two code fields of two bits and three bits, respectively, that together set the mode of operation.

Table 6–4 shows an example of the operational modes along with their codes. Figure 6–17 shows the control field of the U-frame.

Code	Command and responses
00 000	Unnumbered information
00 001	Set normal response mode
00 010	Disconnect mode
00 110	Unnumbered acknowledgment
10 000	Set initialization mode
10 001	Frame reject
11 001	Reset
11 101	Exchange IDs
11 110	Set Asynchronous Balanced Mode Extended

© CENGAGE LEARNING 2012.

Table 6–4: Codes and operational modes in the U-frame

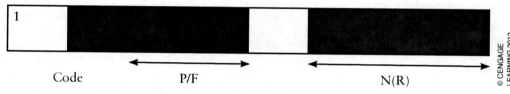

Code   P/F   N(R)

© CENGAGE LEARNING 2012.

Figure 6–17: The frame format of the U-frame

Connection-oriented service is designed to send information and receive an acknowledgment. In this method, the sender initializes the connection, establishes the connection, sends the information, and then waits for acknowledgment. If the acknowledgment is positive and the sender does not have more information to send, then it terminates the connection. Otherwise, the sender will continue to send information until it transmits it all.

Connectionless-oriented service is not a reliable communication since the sender sends its information and hopes the receiver gets the information. There is no acknowledgment either positive or negative.

SUMMARY

The physical connection of computers or other communication devices is called a network topology. A variety of network topology has been developed for various applications. Each network topology has some advantages and disadvantages and a network engineer has to determine which topology is the most suitable for the specific application on which he or she is working. Tree, star, ring, bus, mesh, and hybrid are the most common topologies in a local area network. Computers can be connected to each other individually (point-to-point connection method) or as a group (multipoint connection method). Network systems are set up in either the client-server format or peer-to-peer format.

Protocols are a set of rules and regulations that oversee all operations within a network. This is used so that two communication devices can exchange their information by understanding and accepting the way each side forms and processes information and how that information is transmitted. There are many different protocols for different applications. The Open System Interconnection (OSI) is a seven-layer protocol that allows a variety of existing communication devices to exchange information in a network system. Each layer has its own set of rules that monitor data exchanges between networks to produce a reliable and well-functioning network system. The upper layers, which consist of the top three layers, are responsible for application issues and are used to implement software. The first three layers are known as the media layers and the last four layers are known as the host layers. The seven layers of the OSI start with the physical layer that oversees the physical connection between networks and signaling. The other layers, starting from the second layer are data link, network, transport, session, presentation, and application layers. Industrial data communication is based on the OSI model with only three layers: the physical, data link, and application layers.

To make sure all communication devices in network systems that are manufactured by different vendors are able to communicate together, IEEE has developed a series of standards for all types of network systems. These standards are commonly known as the IEEE 802 standards. The IEEE standards divide the data link layer into two sublayers, the Media Access Control (MAC) layer and the Logical Link Control (LLC) layer. The MAC layer defines the methods that a computer can use to connect to the network via a virtual point-to-point connection. The LLC layer provides connection service between networks that run under different protocols. The other functions of the LLC layer are frame formatting, recognition of the source and destination address, flow control, and error control. Both MAC and LLC layers have their own specific frame formats.

Review Questions

Questions

1. Which network topology is the most reliable?

2. Which topology is the least reliable?

3. Which topology uses a hub as the centerpiece of the network?

4. Which topology uses a repeater to connect the computer to the network?

5. Write the list of rules that a protocol usually sets to establish communication between terminals. What are the differences between the client-server and peer-to-peer network systems?

6. What is the main difference between point-to-point and multipoint connections?

7. The TCP/IP is a protocol that is used in the _____.

8. The first three layers of the OSI model are: _____, _____, and _____.

9. Frame synchronization and formatting are defined by _____.

10. The network layer sets up _____ and provides _____ and _____.

11. The format of transmitting data is defined by _____.

12. What layer determines the method of communication?

13. What is the PDU?

14. What are the header and trailer?

15. What is the difference between the OSI model in ordinary data communication and industrial data communication?

16. What are the two major protocols for the DLL?

17. Draw the HDLC frame format and specify the function of each field.

18. The control field determines the type of information field such as: _____, _____, and _____.

19. What is the purpose of the P/F field in the I-frame, S-frame, and U-frame?

20. The IEEE 802 standard divides the data link layer into _____ sublayers. Name them.

Continues on next page

21. According to the IEEE 802 standards, which sublayer defines the method of accessing the network?

22. Which method of accessing networks is used in Ethernet?

23. What are the functions of the LLC layer?

24. What is the function of the DSAP field in the LLC layer?

25. Draw the LLC frame format.

26. How many types of S-frames are there in terms of commands and responses?

27. What frames are in the control field of the LLC layer? Name all of them.

28. What are the differences between connection and connectionless services?

29. What is piggybacking?

Problems

1. How many cables are needed to implement a mesh topology network with 18 computers?

2. How many computers are arranged in a mesh network topology with 600 cables?

3. Design a network system with three star networks on the existing ring network. The ring network has five computers and each of the star networks has four computers.

4. Design a network system with two ring and one bus networks on the existing star network. The star network has four computers and each of the ring and bus networks has three computers.

5. Design a network system with two ring and one bus networks on the existing star network. The star network has four computers and each of the ring and bus networks has three computers.

IEEE 802.3 Standard Ethernet

Objectives

After completing this chapter, students should be able to:

Discuss the IEEE 802 standards.

Describe the different types of Ethernets.

Describe bridge and switch Ethernets.

Discuss fast and gigabit Ethernets.

As discussed in Chapter 6, IEEE was assigned to develop an Ethernet version I or DIX standard for networking systems that were already in place at Digital, Intel, and Xerox Corporations. DIX Ethernet has a data rate of 10 Mbps with 48 bits of source and destination addressing size. Because the IEEE working group to develop the standard was formed in February (the second month of the year) 1980, the project to standardize computer network technology for the local area networks (LANs) was named IEEE 802 standards. The IEEE 802.3 standard uses the first two physical layers of the OSI model to define the wiring and signaling, the data link layer for addressing, and the multiple access control layer for Ethernet.

7.1 ETHERNET TYPES

The physical layer of the OSI model determines the network's wiring (shielded and unshielded twisted pair, coaxial, or optical fiber) and signaling (frequency of operation, voltage level or optical level, and signal duration) types. The data link layer of the OSI model defines the data frame synchronization, error and flow control, addressing, and how a station can access the network. Ethernet is divided into seven different major types based on the wiring, signaling, and maximum segment length of the network. Each major type may also be subdivided based on their physical link types. The seven types of Ethernet are: Thin Ethernet (10Base2), Thick Ethernet (10Base5), Twisted-Pair Ethernet (10Base-T), Fast Ethernet (100BaseX), Gigabit Ethernet (1000BaseX), 10 Gigabit Ethernet (10GBaseX), and Fiber Ethernet (10Base-F). The first three types are categorized under 10-Mbps Ethernet.

10-Mbps Ethernet

The original 10-Mbps Ethernet standard was developed based on the DIX Ethernet. Later, varieties of the 10-Mbps Ethernet such as 10Base-F and 10Base-T were implemented under the common name of 10BaseX. The 10Base means 10 Mbps using baseband (where the entire capacity of the medium is devoted to only one communication channel). The letter X indicates the length or type of the medium. All 10BaseX Ethernets have the same physical configurations, MAC frame format, and encoding scheme. This chapter will address the following 10BaseX implementations: 10Base2, 10Base5, 10Base-T, and 10Base-F.

The Physical Configurations

The 10-Mbps Ethernet is the earlier version of Ethernet. The 10-Mbps Ethernet LAN was developed to establish data exchanges between two network nodes in a serial-bit format over two unshielded twisted-pair cables. The two node networks, which are known as data terminal equipment (DTE), can be connected to each other directly or via one or more intermediate devices, which are known as data communication equipment (DCE). Computers, servers, or printers are some examples of

DTE devices. Routers, switches, and repeaters are some well-known examples of DCE devices.

Data exchange between a source and destination is possible only if their networks are compatible with each other. The physical layer of the OSI model plays a major part in finding out if two networks are compatible. It defines the protocol and procedures of the bit transmission that includes cable and cable interfaces, connectors, pinouts, and data encoding. To figure out if the two networks are compatible, the physical layer is divided into two sublayers: the physical media independent (PMI) and the physical media dependent (PMD).

The functions of the PMI sublayer include: signaling, data computing, encoding (at the transmitter side) and decoding (at the receiver side), and adding the preamble and start frame delimiter fields into the IEEE 802.3 MAC frame. The PMI sublayer connects to the PMD sublayer by the medium independent interface (MII).

The PMD sublayer itself is divided into two sublayers: the physical coding sublayer (PCS) and the physical medium attachment (PMA). There is also an optional auto-negotiation sublayer. The PCS sublayer provides encoding and multiplexing (at the transmitter side), symbol code alignment and synchronization, and decoding and demultiplexing (at the receiver side). The PMI sublayer, or transceiver, provides signal transmitters/receivers and clock recovery for the received data. The PMD sublayer connects to the physical link (medium) by the medium dependent interface (MDI). Figure 7–1 shows the physical configuration of the 10-Mbps Ethernet.

The Media Access Control (MAC) Frame Format

The MAC layer accepts the transmitted frames from the LLC layer and is responsible for two main tasks: (1) data encapsulation, which means the formatting or assembly of the received frames from LLC, as well as frame transmission and frame error detection; and (2) initialization of the data transmission and data recovery if data transmission fails. To perform these tasks, the MAC frames include the following fields:

1. Preamble is a seven-byte field with alternating 1 and 0 bits (repetition of eight bits of 10101010) to initialize and synchronize data transmission to inform the receiver station that data is coming.
2. Start Frame Delimiter (SFD) is a one-byte field, which is always set to the sequence of 10101011. The last two bits (11) inform the receiver station about the start of the transmitted frame.
3. Destination Address (DA) is a six-byte field that identifies the address of the specific receiver station. The most significant bit of the DA is the left-most bit and it indicates whether the receiver is an individual station (unicast) or a group of stations (multicast). The most significant bit for a unicast address is 0 and 1 for a multicast address. The DA is a broadcast address if all bits of the DA are 1. The DA is universally or locally

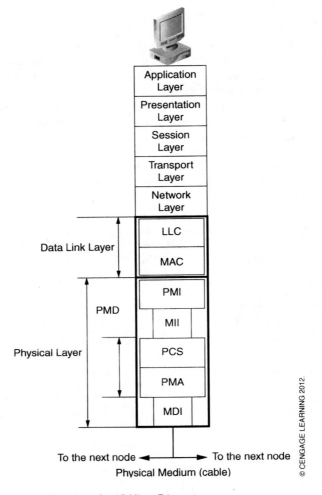

Figure 7–1: The physical configuration of a 10-Mbps Ethernet

administered if the second bit from the left is 0 or 1, respectively. All other remaining 46 bits are a unique code that is assigned to each individual Network Interface Card of the receiver station.

4. Source Address (SA) is a unique six-byte field that is assigned to each individual source station with the most significant bit of 0.

5. Length is a two-byte field that represents the number of bytes in the LLC data field.

6. Data field (the LLC data field) contains a minimum of 46 bytes and a maximum of 1500 bytes of information depending on the type of frame.

7. Pad field has a variable size to complement the data field to ensure there are at least 46 bytes in the data field. If the data field has at least 46 bytes, then the pad field will be empty.

8. Frame Check Sequence (FCS) is a four-byte field that uses the 32-CRC error detection method to detect any possible errors in the received frames (such as damaged or corrupted frames). The FCS field uses the DA, SA, length, and data fields.

Table 7–1 shows the MAC frame format of the 10-Mbps Ethernet.

7 Bytes	1 Byte	6 Bytes	6 Bytes	2 Bytes	0–1500 Bytes	0–46 Bytes	4 Bytes
Preamble	SFD	DA	SA	Length	Data	Pad	FCS

$\longleftrightarrow$
46–1500 Bytes

© CENGAGE LEARNING 2012.

Table 7–1: MAC frame format of the 10-Mbps Ethernet

To ensure receivers can prepare themselves to receive frames in a sequence and without frame interference problems, an idle time between frames was introduced in the Ethernet frame transmission process. The idle time is commonly called Interframe Gap (IFG) and the minimum IFG time in the 10-Mbps Ethernet is 9.6 μsec.

The Access Method

The access method of the 10BaseX Ethernet is the Carrier Sense Multiple Access/ Collision Detect (CSMA/CD) method. All stations in a 10BaseX Ethernet have equal opportunity to share a common communication medium in a switchless environment without an assigned time slot or a primary (central) station. The CSMA/CD access method protocol was developed for this type of environment. In the CSMA/CD access method, each station will sense the status of the common channel. If the channel is in the idle state (no traffic in the channel or there is an interframe gap between frame transmissions), the station will transmit its frames. If the channel is busy (there is traffic in the channel or the channel is in an active status), the station will continue to sense the status of the channel until it becomes idle.

If two or more stations try to send their frames at the same time, there will be a frame collision that causes the frames to become unreadable. If a station detects a collision, it will send a "jam signal" to notify all stations to stop sending their frames for a preassigned waiting period that is determined by a "back-off" algorithm.

A frame collision will also happen if one station sends its frames and the second station starts to send its frame before the completion of the first station's frame transmission. In this situation, the second station will detect the collision immediately but the first station will not detect the collision until the damaged frames propagate back to the first station. To resolve this situation and ensure all collisions will be detected, the maximum network diameter was chosen to be about 2.5 km and the minimum frame length (the minimum of the data frame size) was calculated accordingly.

The propagation time (t_p) is equal to the ratio of the channel length and the propagation velocity in the channel.

$$t_p = \frac{L}{V} \qquad (7.1)$$

L is the length of the channel in meters (or feet). V is the propagation velocity in the channel and is equal to: $V = c/n$ where c is the velocity of the light in the air and n is the channel index of reflection. For example, the index of reflection of a coaxial cable is 1.3. Therefore, the propagation velocity in a coaxial cable is 2.3×10^8 m/sec and the propagation time in a 2.5-km coaxial cable would be $(2500 \text{ m})/2.3 \times 10^8$ m/sec = 10.87 μsec.

It will take $2(10.87 \text{ μsec}) = 21.74$ μsec for the damaged frame to propagate back to the transmitted station.

The transmission time for a data frame (t_{df}) is equal to the ratio of the data frame size and the data rate.

$$t_{df} = \frac{\text{data frame size}}{\text{data rate}} \qquad (7.2)$$

For example, the transmission time for a data frame of 128 bytes in a 10BaseX Ethernet is equal to $t_{df} = (128 \times 8 \text{ bits})/(10 \text{ Mbps}) = 102.4$ μsec, which is much higher than the propagation time of the damaged frame, if the transmission line is a 2.5 km long coaxial cable, (21.74 μsec).

Figure 7–2 shows the mechanism of the CSMA/CD in a flowchart format.

The Encoding Method

The pulses that carry data from the transmitter may lose their power/size (attenuation) or their shape (distortion). The changed pulses have to be regenerated before sending them to the MAC sublayer of the receiver. Filters and pulse-shaping circuits are used to regenerate the changed pulses. The preamble field in the Ethernet MAC frame helps with the synchronization between the transmitter and the receiver. However, a long string of bit 0 in the data field may eliminate the preamble rule in clock recovery and synchronization. An encoding method with an embedded clock for synchronization could help to resolve these problems and reduce the possible transmission errors. To solve these issues, the Manchester encoding method was implemented in most 10-Mbps Ethernets. While the Manchester encoding method is useful in most 10-Mbps Ethernets, it is not a good choice for Ethernets with higher data rates.

The Thin or 10Base2 Ethernet

The 10Base2 Ethernet is a thin coaxial (RG58) 10-Mbps Ethernet that uses a bus topology and operates on baseband bandwidth with a maximum segment length of

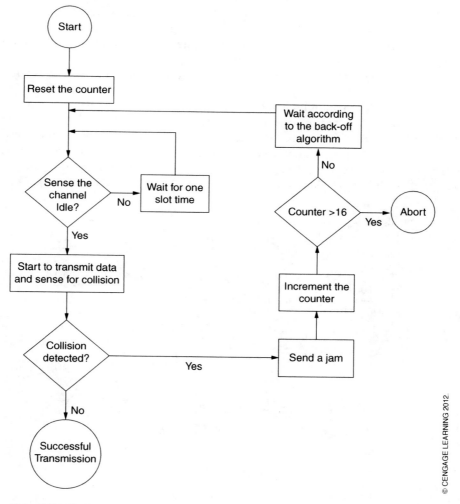

Figure 7–2: CMSA/CD flowchart

200 meters. The Thin Ethernet (Thinnet) is a low-cost LAN network and for this reason it is also called the *cheapnet*. The specifications of the 10Base2 Ethernet network are as follows:

1. Uses the RG58 type of coaxial cable with attached BNC (Bayonet Nut Connector) connectors.
2. The BNC connectors will connect to a T-connector and the T-connector will connect directly to a node. (Do not connect the BNC of the coaxial cable to the BNC of the NIC without a T-connector and do not use a drop cable for connecting a T-connector to a node.)

3. The maximum length of a segment is actually 185 meters.
4. Two 50-Ω terminators are needed at each end of a segment, with one of the terminators being grounded. Terminators are used to prevent signal reflection in the cable.
5. The maximum number of nodes in one segment is 30 (including terminators).
6. Minimum length between two nodes is 0.5 meters (50 cm).
7. The transceiver can optionally be built in the NIC of the node.
8. To extend the number of nodes in the network, two or more segments must be connected to each other via a multichannel repeater.
9. 10Base2 supports only the half-duplex mode of operation.

Figure 7–3 shows a typical 10Base2 Ethernet LAN with two segments.

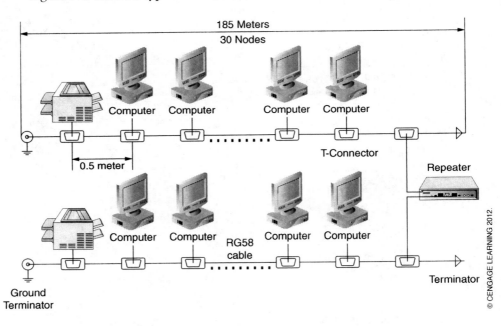

© CENGAGE LEARNING 2012.

Figure 7–3: Thin (10Base2) Ethernet LAN

The Thick or 10Base5 Ethernet

The 10Base5 Ethernet is a thick coaxial (RG8) 10-Mbps Ethernet that uses a bus topology and operates on baseband bandwidth with a maximum segment length of 500 meters. 10Base5 is also called *Thicknet* or *Thickwire Ethernet*. The specifications of the 10Base5 Ethernet network are as follows:

1. Uses the RG8 type of coaxial cable (thick cable).
2. The maximum length of a segment (without repeaters) is 500 meters.

3. Two 50-Ω terminators are needed at each end of a segment, with one of the terminators being grounded. Terminators are used to prevent signal reflection in the cable.

4. The external transceivers are connected to the cable (backbone) via N-connectors. (T-connectors are not allowed.)

5. Nodes are connected to the backbone of the network through their Attached Unit Interface, AUI (a DB-15 socket) via the external transceivers.

6. The maximum number of nodes in one segment is 100 (including terminators).

7. Minimum length between two nodes is 2.5 meters; this helps ensure the reflections from multiple nodes are not in-phase.

8. To extend the number of nodes in the network, two or more segments must be connected to each other via a multichannel repeater (maximum 4-channel repeater).

9. Maximum total length via repeaters is 2500 meters (about 1.56 miles).

10. 10Base5 only supports the half-duplex mode of operation.

11. New nodes can be easily added to the network.

12. Installation of the 10Base5 network is difficult due to the inflexible character of the thick cable.

Figure 7–4 shows a typical 10Base5 LAN with only one segment.

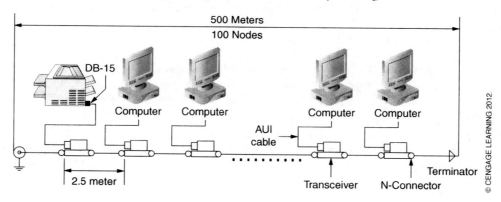

Figure 7–4: Thick (10Base5) Ethernet LAN

The 10Base-T Ethernet

The 10Base-T Ethernet uses unshielded twisted-pair (UTP) category-3 cable, which is similar to a telephone wire. It actually uses only two twisted-pair wires and for this reason, it is called *10Base-Twisted pair* or simply *10Base-T*. The twisted-pair cables have the capability to reduce radio frequency interference. The transmitter can send signals with two different voltages: +2.5 V or −2.5 V. The 10Base-T Ethernet operates over a distance of up to 100 meters and uses a star topology in which the nodes are connected to an active hub (or switch) as a central station. Each segment, including the hub and nodes, can be connected to other segments in a daisy-chain fashion. If collisions occur in the 10Base-T, they happen in the hub not in the cable as in 10Base2 and 10Base5. 10Base-T supports both half-duplex and full-duplex and it also operates on baseband bandwidth. The specifications of the 10Base-T Ethernet network are as follows:

1. Uses two pairs of the four-pair UTP category-3 cables.
2. Maximum length of a segment is up to 100 meters.
3. Nodes are connected to the hub or switch by an 8-pin RJ-45, Registered-Jack (older connector type) or 8P8C (the new connector type). RJ-45 or 8P8C connectors are used as twisted-pair wire terminators.
4. All nodes share the 10 Mbps when a hub is used as a central station (a shared media LAN).
5. Maximum of 1024 segments in a network.
6. Maximum of two nodes per segment.
7. Minimum distance of 3 meters between nodes.
8. Hub acts as a central station and up to four daisy-chained hubs can be used.
9. Uses a star topology.
10. 100-Ω impedance.

The RJ-45 connector has eight pins and four pairs of wires. Pairs of wires are connected to the eight pins as follows:

1. Pair-1 (blue to pin-4 and white/blue to pin-5)
2. Pair-2 (orange to pin-2 and white/orange to pin-1)
3. Pair-3 (green to pin-6 and white/green to pin-3)
4. Pair-4 (brown to pin-8 and white/brown to pin-7)

Two out of the four twisted-pair wires must be selected before connecting a node to the hub via an RJ-45 connector.

The 8P8C (8-position, 8-contact) connector is in fact not truly compatible with the RJ-45, even though it is often called an RJ-45. The 8P8C connector has the two parts of a male plug and a female jack. Figure 7–5 shows a typical 10Base-T LAN.

The 10Base-Fiber Link

The 10Base-Fiber Link (10Base-FL) Ethernet uses two pairs of multimode optical fiber cables to connect nodes to a 10Base-FL repeater via a point-to-point interface. One of the multimode optical fiber pairs is used to transmit, and the other is used to receive data. Each segment of the 10Base-FL has two attached transceivers with one at each end of the segment. The Straight Tip connector (ST-connector) is used to attach transceivers to the multimode optical fiber. The 62.5/125 multimode optical fiber is the most common type used in 10Base-FL in order to achieve the 2-km

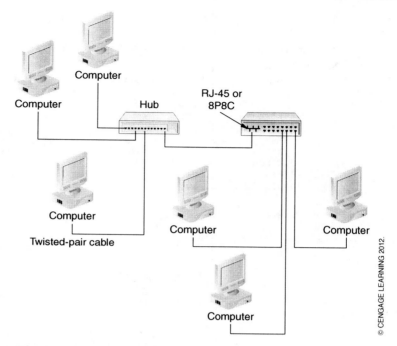

Figure 7–5: 10Base-T Ethernet LAN

maximum segment length. This type of optical cable has a 62.5-μm core diameter and a 125-μm external diameter and can transmit a light signal with a wavelength of 850 nm. The 10Base-FL is actually a modified version of the Fiber Optic Inter-Repeater Link (FOIRL) standard. FOIRL had also supported a 10-Mbps data rate over a 1-km point-to-point communication between two repeaters. However, the 10Base-FL standard has extended the maximum communication link of the FOIRL to 2 km, and has made communication possible between two repeaters and two computers, or one repeater and one computer.

The advancements in technology that have helped design and produce the high quality of optical fibers and transceivers, as well as the independent nature of transceivers in the 10Base-FL, have both allowed the use of full-duplex operation. These have also helped extend the maximum length of the segment to more than 5 km. The summary of the 10Base-FL specifications are:

1. Transmission rate: 10 Mbps
2. Maximum segment length: 2000 m (about 1.25 miles)
3. Transmission media: Two 62.5/125 multimode optical fibers
4. Signal type: Light signal with λ = 850 nm
5. Connector type: Straight Tip (ST)

6. Transceiver type: External with a maximum of two transceivers per segment
7. Network topology type: Star with repeater as a core (primary) station
8. Signal encoding type: Manchester

Figure 7–6 shows a typical 10Base-T LAN.

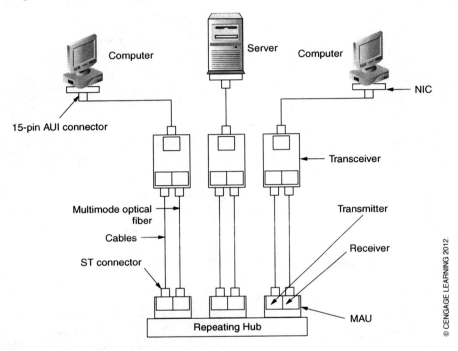

Figure 7–6: 10Base-FL Ethernet LAN

The extension of the 10Base-FL operational length is possible by increasing the number of repeaters in the network and optimizing the point-to-point communication between repeaters. But there are some problems that are associated with repeaters. Repeaters add signal delay that will cause the collision propagation time to exceed the allowable limit and reduce the preassigned interframe gap time (9.6 μsec) in a 10-Mbps network. The 10Base-Fiber Backbone, or simply 10Base-FB, was developed to increase the number of repeaters in the 10Base-FL while maintaining the allowable interframe gap time. The 10Base-FB operates on a special synchronous signaling link that establishes a synchronized transmission between repeaters. As a result, the interframe gap time shrinks by a factor of 4. Unlike 10Base-FL, the 10Base-FB cannot support the full-duplex mode of communication.

The 10Broad36 is the only 10-Mbps Ethernet that does not operate on baseband. The 10Broad36 is able to transmit signals over 3600 m via a 75-Ω cable TV (CATV) broadband cable. Unlike the baseband transmission, the signal flow in the broadband transmission is unidirectional, which means signals travel from one station to another station in a single path and is retransmitted in another path. The 10Broad36, however, has become obsolete with the development of fiber-based networks.

7.2 HALF- AND FULL-DUPLEX ETHERNET

The half-duplex mode of operation supports either transmission or reception of data frames; that is, no station can transmit and receive data frames simultaneously. Ethernet is the only mode of communication (operation) in the 10Base2 and the 10Base5 Ethernet networks. It is also used in the 10Base-T and the 10Base-FL Ethernet networks, but both of these are capable of supporting the full-duplex mode of operation. Half-duplex Ethernet uses the CSMA/CD type of multiple access protocol. As discussed in the Access Method section, there is a minimum time requirement for completing the transmission of a frame and detecting collisions. This minimum time must be less than one "slot time" or time duration of 512 bits. The slot time is a crucial parameter in the half-duplex operation. It defines the maximum segment length, number of repeaters in each path, minimum size of the Ethernet frame, and the detection time of a collision (if the collision takes place, it must be detected within the first 512 bits).

The full-duplex mode of operation supports the simultaneous transmission and reception of data frames in a point-to-point connection method over separate paths without interference. As a result, there would be no collision when two stations are communicating with each other. Therefore, there is no need for the CSMA/CD multiple access protocol in full-duplex mode. Consequently the segment length will not be limited based on slot time as it is in the half-duplex mode of operation.

The other advantages of the full-duplex mode of operation are in efficient use of the link and doubling of the throughput size.

7.3 BRIDGE AND SWITCHED ETHERNET

Sharing the bandwidth of the network and collision domain are the two major issues in 10-Mbps shared networks. The bridge and switched networks have helped network engineers and administrators to enhance the 10-Mbps networks while the speed (data rate) still remains an issue in the growing demand for fast Ethernet.

In bridge Ethernet, the entire network is divided into two smaller segments with almost an equal number of nodes. The divided segments will then be connected to each other via a bridge. Segmentation of the network and connecting the individual segments via a bridge will increase the network length (acts like a repeater) and regulate traffic. A bridge operates in the first two layers of the OSI model (physical and data link layers).

In the bridge network, the sender transmits frames. The bridge examines the address of the destination address. If the destination and source are located in the same segment, then there is no action that is taken by the bridge. In this case, the bridge acts like a filter and it allows two other nodes in the same segment to communicate with each other simultaneously. If the destination and source are located in different segments, then the bridge will forward the frame to the designated segment. If transmission and reception of the frame take place over two bridges, then a special frame, called a *bridge frame,* will be created to facilitate the communication. Figure 7–7 shows a typical bridge network.

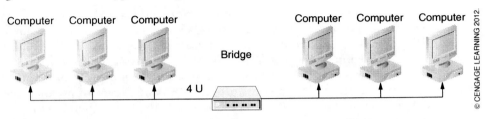

Figure 7–7: Typical bridge network

Switched Ethernet also divides the network, but it divides it into multiple segments. There are two types of switches: two-layer and three-layer switches. A two-layer switch operates at the physical and data link layers. In this sense, the two-layer switch functions like a high-speed bridge. The three-layer switch operates at the network layer and functions like a high-speed router. In switched Ethernet, each segment contains an end station and the switch. Therefore, the frame packets will be forwarded to the designated destination. This mechanism will create many simultaneous communications in the network. The backbone Ethernet switch is capable of supporting hundreds of segments.

Transmission of frame packets in switched Ethernet takes place as follows:
1. The switch accepts every transmission.
2. It reads the address of the destination that is written in the header of the transmitted packet.
3. It will find the location of the destination from the destination address table.
4. It will establish a temporary multiple crossover connection between the source and destination.
5. It will forward the frame packets to the designated destination and then terminate the connection after communication is completed.

Figure 7–8 shows a typical switched Ethernet network.

The full-duplex switched network has the capability of increasing the data rate of an Ethernet network by a factor of 2. For example if the full-duplex is used in those 10-Mbps Ethernet networks that use half-duplex, such as 10Base2 or 10Base5, the data rate will be increased from 10 Mbps to 20 Mbps because the full-duplex establishes two independent communication paths, one for transmission and one for reception of data simultaneously. Figure 7–9 shows a full-duplex switched Ethernet network.

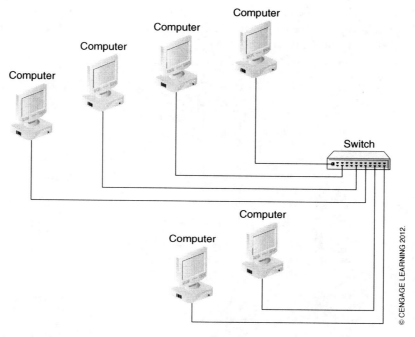

Figure 7–8: Typical switched Ethernet network

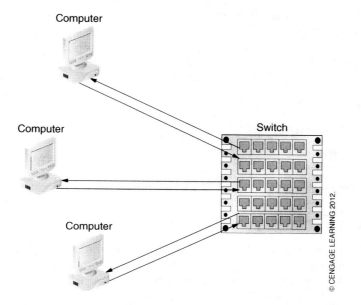

Figure 7–9: Typical full-duplex switched Ethernet network

The switch can also to be used as a backbone of the network to connect different types of networks with different capacities. Figure 7–10 shows a typical backbone switched Ethernet network.

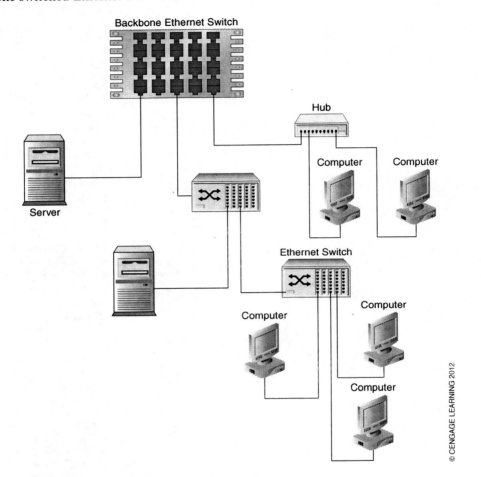

Figure 7–10: Typical backbone switched Ethernet network

7.4 FAST ETHERNET

Fast Ethernet was developed to provide Ethernet with a higher data rate but retain compatibility with the existing 10-Mbps Ethernet standard and existing network installations. The Fast Ethernet or 100-Mbps Ethernet standard (IEEE 802.3u) kept the frame length, frame format, and address size of the 10-Mbps standard (IEEE 802.3). Fast Ethernet adopted the point-to-point and the star network topology with a switch or hub as a central node of the network. The star topology supports both half-duplex and full-duplex modes of operation. The half-duplex uses the CSMA/CD protocol and the full-duplex does not. Therefore, the Fast Ethernet has implemented the CSMA/CD in its standard for the cases when the half-duplex mode is the choice of communication.

Unlike the IEEE 802.3 standard, Fast Ethernet employs different encoding methods depending on the type of implementation. For example, while the 100Base-TX employs the 4B/5B block coding that fed into the MLT-3 encoding method, the 100Base-FX employs the NRZ-I encoding method.

There is an additional sublayer in the physical layer of the Fast Ethernet that is called auto-negotiation. The purpose of this sublayer is to allow a station to sense the mode of operation and data rate of another station and automatically negotiate which has the better mode of operation and highest data rate. For example, auto-negotiation allows the 10Base-T network to communicate with a 100Base-TX network while both networks continue to function at their own designated data rates. The adaptor that can automatically sense the speed of the line and is able to adjust itself to the higher data rate is called the "10/100" adapter. Figure 7–11 shows a typical hybrid network that uses 10Base-T and 100Base-TX networks. Fast Ethernet implementations include: 100Base-T2, 100Base-T4, 100Base-TX, and 100Base-FX.

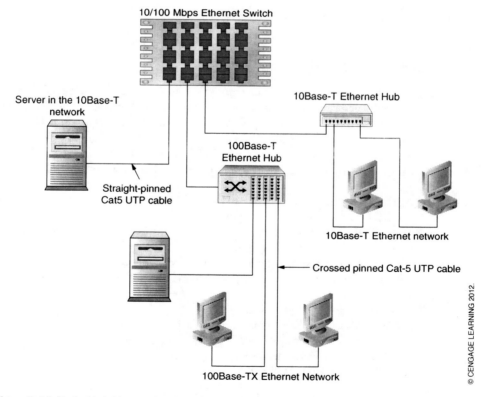

Figure 7–11: Typical hybrid network using 10Base-T and 100Base-TX

© CENGAGE LEARNING 2012.

The 100Base-T2 Fast Ethernet

The 100Base-T2 Fast Ethernet was developed to allow users of 10Base-T to upgrade their network from 10 Mbps to 100 Mbps. The 100Base-T2 transceiver transmits data over two pairs of UTP-3 (EIA/TIA category-3) cable, which is the cable of choice for 10Base-T. It is also able to transmit data simultaneously over the four pairs of UTP-3, with two pairs for the 100Base-T2 and two pairs for 100Base-T2, 10Base-T, or digital phone.

The 100Base-T2 Ethernet uses the "dual-duplex" baseband transmission of data using the 4D-PAM5 encoding method that allows transmitting 4 bits of data per signal transition on each pair of wires. This translates to a 25-MBaud modulation rate.

The 100Base-T2 specifications are:
1. Transmission rate: 100 Mbps (200 Mbps if full-duplex is employed)
2. Cable type: 100-Ω impedance rating UTP-3 (two-pair) cable (also supports use of a four-pair UTP-3 cable)
3. Connector type: 8P8C (8-position 8-contact)
4. Maximum segment length of cable: 100 meters
5. Maximum number of transceivers per segment: 2
6. Network topology: Star with point-to-point connection
7. Encoding method: 4D-PAM5

The 100Base-T4 Fast Ethernet

The 100Base-T4 Fast Ethernet transmits data over a four-pair UTP-3, UTP-4, or UTP-5 cable. Six of the eight wires are used for transmission and reception of data, and the remaining two wires are used for collision detection. The 100Base-T4 uses the 8B6T encoding method that allows 8 bits of data to be encoded in a 6-ternary signal transition over three twisted-pair wires. Therefore, the 100-Mbps data transmission rate is divided in three and each of the twisted pairs has a data rate of $100/3 = 33.33$ Mbps.

The 100Base-T4 has its own specific transceiver, which is either connected to the MII connector or is embedded in the Network Interface Card (NIC). The 100Base-T4 Ethernet does not support a full-duplex mode of operation.

Specifications for the 100Base-T4 are:
1. Transmission rate: 100 Mbps (does not support full-duplex)
2. Cable type: 100-Ω impedance rating UTP-3, UTP-4, or UTP-5 (four pairs) cable
3. Connector type: 8P8C
4. Maximum segment length of cable: 100 meters
5. Maximum number of transceivers per segment: 2
6. Network topology: Star with point-to-point connection
7. Encoding method: 8B6T

The 100Base-TX Fast Ethernet

The 100Base-TX Fast Ethernet transmits data over two pairs of UTP-5 cable with one pair for transmission and one pair for reception of data. The UTP-5 cable is used along with an 8P8C connector and EIA/TIA 568B pinning devices. The 100Base-TX also supports the 150-Ω shielded twisted-pair (STP) cable with a DB9 connector that is used in Token Ring wiring. If the 100Base-TX repeating hub and the station's NIC have embedded transceivers, then the station can be connected directly to the repeating hub by UTP-5 cable and 8P8C connectors. If the station does not have an embedded transceiver, it must be connected to its MII connector. Unlike the 100Base-T4, the 100Base-TX supports the full-duplex mode of operation.

The MLT-3 encoding method has been used in the 100Base-TX Fast Ethernet. Since the MLT-3 is not a self-synchronous encoding method, the encoding process starts with 4B/5B block coding to provide a data rate of 125 MBaud with a net data rate of 100 Mbps to add the synchronization feature to the MLT-3 encoding. The 100Base-T4 specifications are:

1. Transmission rate: 100 Mbps (200 Mbps if full-duplex is employed)
2. Cable type: 100-Ω impedance rating UTP-5 (two pairs) cable (150-Ω STP can also be used)
3. Connector type: 8P8C (DB9 if the STP cable is used)
4. Maximum segment length of cable: 100 meters
5. Maximum number of transceivers per segment: 2
6. Network topology: Star with point-to-point connection
7. Encoding method: 4B/5B block coding (to provide synchronization) and MLT-3 (for encoding/decoding)

The 100Base-FX Fast Ethernet

The 100Base-FX Fast Ethernet transmits data over two multimode optical fiber cables, one for transmission and one for reception. Therefore, the 100Base-FX uses the full-duplex mode of operation. 100Base-FX also supports single-mode optical fiber and half-duplex with smaller segment length. The maximum segment length for the multimode optical fiber cable along with full-duplex is 2 km and with half-duplex is 412 meters. If the single-mode cable and full-duplex are used, the maximum segment length can be extended to 10 km. The maximum segment length from a station to the repeating hub is 150 meters. The recommended connector for the 100Base-FX is SC, but ST and MIC connectors also can be used. The connectors and transceiver are usually embedded in the repeating hub and NIC that connects the optical fiber cable directly from a station to the repeating hub.

The multimode optical fiber cable in the 100Base-FX is usually a 62.5/125 type, as is used in the 10Base-F network but the light signal wavelength is 1300 nm.

The 4B/5B block coding in conjunction with NRZ-I are used for encoding the signal in the 100Base-FX. The 100Base-FX specifications are:

1. Transmission rate: 100 Mbps (200 Mbps if full-duplex is employed)
2. Cable type: Two multimode optical fiber cables (single mode can also be used)
3. Connector type: SC (recommended), ST, and MIC
4. Maximum segment length of cable: 2 km (if multimode optical fiber cable and full-duplex are used) and 412 meters (if half-duplex is used)
5. Maximum number of transceivers per segment: 2
6. Network topology: Star with point-to-point connection
7. Encoding method: 4B/5B block coding (to provide synchronization) and NRZ-I (for encoding/decoding)

7.5 GIGABIT ETHERNET

Gigabit Ethernet (IEEE 802.3z and IEEE 802.3ab) was developed to increase the data rate of the Fast Ethernet by 10 times (100 Mbps to 1000 Mbps or 1 Gbps) while remaining compatible with the Fast Ethernet with the same frame size, same frame format, and same limitation for the frame length. It was also intended to support the auto-negotiation characteristic of the Fast Ethernet. To accomplish these tasks, two separate standards; IEEE 802.3 and ANSI X3T11 Fiber Channel (Fiber Channel is a high-speed data transfer interface that is able to transfer a large volume of information very fast) were integrated as follows:

1. The LLC layer is the same as the IEEE 802.2 standard.
2. The MAC layer includes half-duplex (using CSMA/CD) and full-duplex.
3. The physical layer is divided similar to the Fast Ethernet with the following functions and specifications:
   a. The Physical Coding Sublayer (PCS) supports the 8B/10B encoding/decoding method and auto-negotiation function. The IEEE 802.3ab uses the 4D-PAM5 encoding method.
   b. The Physical Medium Attachment (PMA) sublayer does serialization/deserialization (serialize the incoming bits from PCS and forward them to the lower sublayer [PMD] or deserialize the incoming bits and forward them to the PCS sublayer).
   c. The Physical Medium Dependent sublayer defines the specifics of the Gigabit Ethernet implementation (type of transceivers and cables).
4. The Medium Independent Interface (MII) of the Fast Ethernet is replaced by the Gigabit Medium Independent Interface (GMII). The GMII can operate at both 10 Mbps and 100 Mbps, and 8 bits (one byte) of data are moved across the GMII per clock cycle.

Increasing the data rate comes with decreasing the bit length or slot time. The Gigabit Ethernet has a slot time of 512 μsec, which results in reduction of the collision time. Consequently the maximum segment length in the Gigabit Ethernet is reduced to 25 meters, which is not an appropriate length. The Carrier extension (use to increase the minimum frame length) and frame bursting (which helps to transmit multiple frames) mechanisms are used to increase the segment length of the Gigabit Ethernet.

Figure 7–12 shows the Gigabit Ethernet specifications.

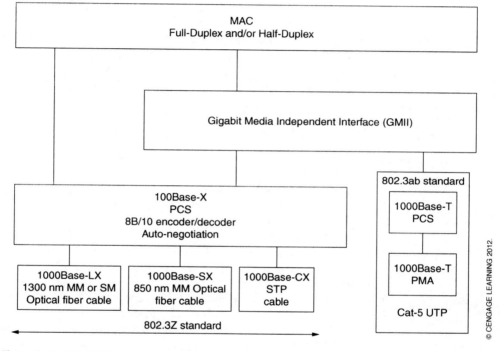

Figure 7–12: Gigabit Ethernet specifications

The Gigabit Ethernet uses the Gigabit Interface Converter (GBIC). The GBIC helps to arrange each of the gigabit ports on a port-by-port basis for short wavelength (SX in the 1000Base-SX), and long-wavelength (LX in the 1000Base-LX) interfaces. Both multimode (62.5 μm and 50 μm) and single-mode optical fiber cables are used in the Gigabit Ethernet. Figure 7–13 shows the implementation of the Gigabit Ethernet using GBIC.

The 1000Base-SX

The 1000Base-SX is a Gigabit Ethernet implementation that uses short-wavelength (770 to 860 nm) lasers as a source for its light signal. It transmits data over

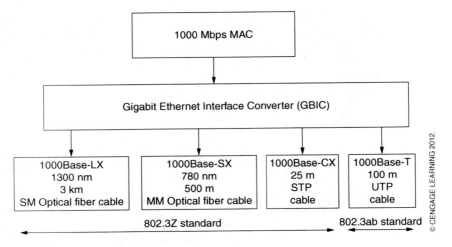

Figure 7–13: Implementation of the Gigabit Ethernet using GBIC

two multimode optical fiber cables. The specifications of the 1000Base-SX are as follows:

1. Transmission rate: 1000 Mbps (2000 Mbps if full-duplex is employed)
2. Cable type: Two multimode optical fiber cables (62.5/125 or 50/125)
3. Signal type: 770 to 860 nm light signal
4. Connector type: Duplex SC
5. Maximum segment length of cable: 316 meters (half-duplex and 50/125 optical fiber cable) and 550 meters (full-duplex and 50/125 optical fiber cable)
6. Maximum number of transceivers per segment: 2
7. Network topology: Star with point-to-point connection
8. Encoding method: 8B/10B block coding (to provide synchronization) and NRZ-I (for encoding/decoding)
9. Minimum output power: 9.5 dBm
10. Minimum receiver sensitivity: 17 dBm

The 1000Base-LX

The 1000Base-LX Gigabit Ethernet implementation uses long-wavelength (1.27 to 1.355 μm—near infrared) lasers as a source for its light signal. It transmits data over two multimode optical fiber cables. The specifications of the 1000Base-LX are as follows:

1. Transmission rate: 1000 Mbps (2000 Mbps if full-duplex is employed)
2. Cable type: Typically two multimode optical fiber cables (62.5/125 or 50/125)
3. Signal type: 1.27 to 1.355 μm light signal

4. Maximum RMS spectrum width: 4 nm
5. Connector type: Duplex SC
6. Maximum segment length of cable: 500 meters (full-duplex and multi-mode 5-μm optical fiber cable) and 5000 meters (full-duplex and 10-μm single-mode optical fiber cables)
7. Maximum number of transceivers per segment: 2
8. Network topology: Star with point-to-point connection
9. Encoding method: 8B/10B block coding (to provide synchronization) and NRZ-I (for encoding/decoding)

The 1000Base-CX

The 1000Base-CX Gigabit Ethernet implementation uses a special balanced shielded twisted-pair copper cable called *twinaxial* or *short-haul copper*. The specifications of the 1000Base-CX are as follows:

1. Transmission rate: 1000 Mbps (2000 Mbps if full-duplex is employed)
2. Cable type: Twinaxial (shielded balanced twisted-pair) cable
3. Connector type: DE-9 (a D-shaped, 9-pin connector which is also called *D-subminiature connector*) or 8P8C (8-pin, 8-contact connector)
4. Maximum segment length of cable: 25 meters
5. Maximum number of transceivers per segment: 2
6. Network topology: Star with point-to-point connection
7. Encoding method: 8B/10B block coding (to provide synchronization) and NRZ-I (for encoding/decoding)

The 1000Base-T

The 1000Base-T is the IEEE 802.3ab standard, while the above three implementations are part of the IEEE 802.3z standard. The 1000Base-T transmits data simultaneously in two directions over four pairs of category-5 balanced copper cables (category-6 may also be used). Transmission goes through echo cancellation and a 4D-PAM5 encoding method. 1000Base-T supports full-duplex and uses two methods of signaling that are used in 100Base-T2 (25 MBaud PAM5) and 100Base-TX (125 MBaud MLT-3). This eases the communication between 1000Base-T and 100Base-TX through 100/1000 dual speed Ethernet adaptors. Auto-negotiation is required for 1000Base-T.

The specifications of the 1000Base-T are as follows:

1. Transmission rate: 1000 Mbps (2000 Mbps if full-duplex is employed)
2. Cable type: Four pairs of category-5 or category-6 balanced copper cables
3. Connector type: 8P8C
4. Maximum segment length of cable: 100 meters
5. Maximum number of transceivers per segment: 2

6. Network topology: Star with point-to-point connection
7. Encoding method: 4D-PAM5

7.6 10GIGABIT ETHERNET

The 10Gigabit Ethernet (10GbE) is the IEEE 802.3ae standard that is fully compatible with other Ethernet standards with lower data rates. To be fully compatible, the 10Gigabit Ethernet uses the same frame size and frame format, as well as the Media Access Control (MAC) protocol. The 10Gigabit Ethernet transmits 10,000 Mbits (1 gigabit) of data per second over optical fiber cables or twisted-pair copper cables. The 10GbE supports only the full-duplex mode of operation and uses the 64B/66B encoding method and a new optical PMD. The implementations of 10GbE are either based on the optical fiber cable (10GBase-SR, 10GBase-ER, 10GBase-LR, 10GBase-SW, 10GBase-LW, and 10GBase-EW) or on the twisted-pair copper cables (10GBase-T and 10GBase-CX). Table 7–2 shows some specifications for 10GbE implementation.

Implementation	Segment length	Wavelength	Cable
10GBase-SR	300 m	850 nm	Multimode (MM) optical fiber
10GBase-ER	40 km	1550 nm	Single-mode (SM) optical fiber
10GBase-LR	10 km	1310 nm	SM optical fiber
10GBase-SW	300 m	850 nm	MM optical fiber
10GBase-LW	10 km	1310 nm	SM optical fiber
10GBase-EW	40 km	1550 nm	SM optical fiber
10GBase-T	100 m	—	UTP
10GBase-CX	15 m	—	Eight pairs, twinaxial, shielded copper cable

Table 7–2: Some specifications for 10GbE implementation

SUMMARY

In 1980, the IEEE was assigned to develop standards for the Ethernet based on the existing 10-Mbps DIX Ethernet. IEEE took the OSI model and changed its data link layer and physical layer to create a new standard for 10-Mbps Ethernet (IEEE 802 standard) so that network devices designed and built by different manufacturers would be able to communicate with each other. In the IEEE 802 standard, the DLL of the OSI model was divided into the LLC and MAC sublayers. The physical layer was also divided into PMI and PMD sublayers. These two sublayers are connected together by MII, and the PMD is connected to the physical link by MDI. The MAC frame consists of preamble, SDF, DA, SA, Data, Pad, and FCS fields. The collision detection method for the half-duplex mode of operation is the CSMA/CD protocol.

10-Mbps Ethernet operates on baseband and is able to transmit 10 megabits of data per second over coaxial, UTP, and optical fiber cables. 10Base2, 10Base5, 10Base-T, and 10Base-FL are different implementations of the 10-Mbps Ethernet with their own characteristics and specifications.

The second generation of the IEEE standard (IEEE 802.3) was developed to be compatible with the 10-Mbps Ethernet but with 10 times faster data rates of 100 Mbps. The Fast Ethernet uses the point-to-point connection and star network topology. The encoding in the Fast Ethernet is different than in the 10-Mbps Ethernet. The Fast Ethernet implementations include: The 100Base-T2, 100Base-T4, 100Base-TX, and 100Base-FX.

The Gigabit Ethernet is the third generation of the IEEE 802 standard (IEEE 802.3z and IEEE 802.3ab) which was developed based on the IEEE 802.3 and the ANSI X3T11 Fiber Channel standards to increase the data rate of the Fast Ethernet by a factor of 10 (100 to 1000 Mbps or 1 Gbps), while remaining compatible with the Fast Ethernet with the same frame size, same frame format, and same limitation for the frame length. The encoding method of Gigabit Ethernet is different than that of the Fast Ethernet. The Gigabit Ethernet implementations include: The 1000Base-SX, 100Base-LX, 100Base-CX, and 100Base-T.

The 10Gigabit Ethernet is the fourth generation of the IEEE 802 standard (IEEE 802.3ae standard) which is fully compatible with the Fast and Gigabit Ethernet networks. To be fully compatible, the 10Gigabit Ethernet uses the same frame size and frame format, as well as the Media Access Control (MAC) protocol.

Review Questions

Questions

1. Which two layers of the OSI model were used to develop the IEEE 802 standard?

2. What is the difference between baseband and broadband communications?

3. Explain what 10Base-T means.

4. Define *data terminal equipment* and *data communication equipment*.

5. What is the function of the PMI sublayer?

6. What is the function of the PMD sublayer?

7. How many bits are in the Preamble field? Write its bit sequence.

8. What are the sizes of the DA and SA fields?

9. How do you recognize if the DA address is a broadcast address?

10. What are the minimum and maximum sizes of the data field?

11. What type of error correction is used in the FCS field?

12. What type of collision detection method is used in 10-Mbps Ethernet?

13. What does IFG stand for and what does it do?

14. What is the maximum segment length of the 10Base2?

15. What is the mode of operation in the 10Base5?

16. What is the network topology in the 10Base-T?

17. What type of transmission medium is used in the 10Base-FL?

18. What types of connectors are used in the 10Base-FL?

19. What type of transmission mode requires CSMA/CD protocol?

20. What are the advantages of the bridge Ethernet over the common 10-Mbps Ethernet?

21. What are the advantages of the switched Ethernet over the common 10-Mbps Ethernet?

22. What are the commonalities between 10-Mbps and 100-Mbps Ethernet networks?

23. What is the encoding method in the 100Base-T2?

24. What is the encoding method in the 100Base-TX?

25. What types of connectors are used in 100Base-FX?

26. What is the function of the PCS sublayer in Gigabit Ethernet?

27. What is the function of the PMA sublayer in Gigabit Ethernet?

28. What is the function of the GBIC in the Gigabit Ethernet?

29. What is the signal type of 1000Base-SX?

30. What is the maximum RMS spectrum width in the 1000Base-LX?

31. What type of block coding is used in 1000Base-CX?

32. What is the maximum segment length of the 1000Base-T?

33. What type of transmission mode is used in 10GbE?

34. What are the commonalities between Gigabit and 10Gigabit Ethernet networks?

35. What type of block coding is used in 10GbE?

Problems

1. What is the propagation time in a 5-km coaxial cable that has a speed of 2.3×10^8 m/sec?

2. What are the speed and the propagation time in a 2-km cable that has an index of reflection of 1.26?

3. What is the transmission time for a data frame of 512 bytes in the 10Base-T?

4. What is the transmission time for a data frame of 512 bytes in the 100Base-T?

8

IEEE 802.5 Standard: Token Ring Local Area Network

Objectives

After completing this chapter, students should be able to:

- *Describe the characteristics of the IEEE 802.5 standard (Token Ring).*

- *Discuss the operation and architecture of Token Ring and Token Bus network systems.*

The Token Ring local area network (LAN) was developed by IBM in the 1970s for its mainframe network architecture. IBM's mainframe network architecture was called *System Network Architecture* (SAN). In 1985, IBM introduced its 4-Mbps Token Ring LAN using unshielded twisted pair (UTP) cable. At this time, the IBM Token Ring, with some minor changes, became the IEEE 802.5 standard. In 1989, IBM introduced its 16-Mbps Token Ring protocol using shielded twisted pair (STP) cable. Consequently, some modifications took place in the IEEE 802.5 standard to become adaptable to the new Token Ring data rate. Some other token rings with data rates such as 10 Mbps and 12 Mbps have been developed by proNet-10 and Apollo Computer but they are not compatible with the IBM Token Ring protocol.

8.1   OPERATION OF THE TOKEN RING LAN

The IBM Token Ring implementation was based on logical ring topology, where data is transmitted from one computer to the next sequentially and the stations (i.e., computers) are physically connected to a central network device (hub) in a star topology. The hub is the actual physical ring of cable.

In the Token Ring LAN, one computer in the ring will be elected as an *active monitor* (AM) and all other computers become *standby monitors* (SMs). Every standby monitor can become an active monitor by requesting and going through the monitor contention process. The monitor contention process takes place when the current AM computer is not functioning and, therefore, there is no AM in the ring or a loss of signal is detected in the ring. The AM computer is responsible for the following tasks:

1. Maintaining the master clock of the ring for synchronizing the signal for all computers in the ring.
2. Maintaining the circulation of the token in the ring by inserting a 3-byte delay in the ring to make sure there is sufficient buffering size in the ring for the token circulation.
3. Detecting a lost token or broken ring.
4. Performing ring polling every 7 seconds and ring purging when detecting a problem within the ring operation. Ring purges reset the ring if a loss of data or interruption in the Token Ring operation is reported.
5. Removing the frames that keep circulating from the ring by looking into the monitor bit (M-bit).
6. Maintaining circulation of only one token if there is no frame transmission in the ring.
7. Sending a beacon frame to all computers in the ring to notify them about transmission lost in the ring.

When none of the computers in the ring has any data frames for transmission, then the token frame will circulate around the ring from one computer to another. When any of the computers has data for transmission, it must get hold of the token frame, change the token frame in the data frame by modifying the token bit (T-bit) from logic 0 to logic 1, and then transmit data in the ring. Token rings provide an optional medium access (token priority) method that allows each computer to prioritize itself by changing the priority bits (P-bit). The eight bits of the access frame consist of the: M-bit, T-bit, and P-bits (3-bit field) along with the R-bits (reserved bits, also a 3-bit field).

The transmitted frame will circulate around the ring from one computer to another. Each computer will read the destination address in the transmitted frame. If it matches with its own address then it will copy the data frame into its buffer, set the frame status field to notify the sender that the data has been received, and finally retransmit the frame into the ring. The retransmitted frame again will be circulated around the ring until it reaches the sender (when the source address and the computer address match each other). The sender will remove the data from the frame and set the T-bit to logic 0 in order to convert the data frame to a token frame. If the source (sender) computer is down, the data frame will circulate around the ring more than one time. In this situation, the active monitor will remove the data frame from the ring.

8.2 TOKEN RING NETWORK ARCHITECTURE

The Token Ring network architecture consists of an active multiple access unit (MAU) or multistation access unit (MSAU) and several computers. Computers are connected to the MAU via a cable. The length of the cable segment can be extended to 10 times longer if an optical fiber is used. The IBM 8228 MAU has ten ports that include eight ports for connecting computers to the ring and two ports that are associated with the ring-in (RI) and ring-out (RO). Figure 8–1 shows a Token Ring network layout.

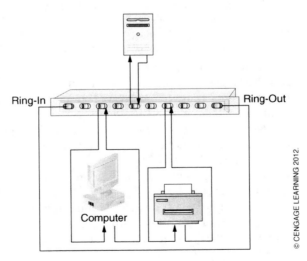

Figure 8–1: A Token Ring network layout

The RI and RO ports are used for the extension of the Token Ring network and construction of the backup ring. The Token Ring network has two rings, main and backup rings, which are parallel to each other. The backup ring is created by connecting the RO of the first MSAU to the RI of the second MSAU or vice versa. The data in the main and backup rings flow in the opposite direction. Existence of the backup ring is an advantage of the Token Ring network by providing a continuous data flow in case of trouble in the ring. Figure 8–2 shows a Token Ring network with two MSAUs and the main and backup rings.

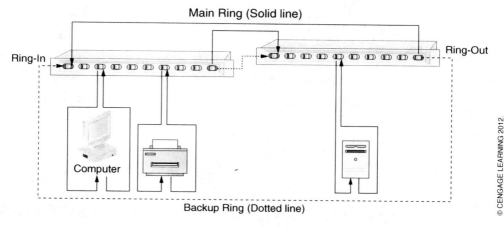

Figure 8–2: A Token Ring network with main and backup rings

Under regular circumstances, when there is no report on trouble in the ring, the Token Ring network uses only the main ring. The backup ring is used only when there is trouble in the ring, such as a broken line. In this situation, the main ring can be "wrapped" to the backup ring, which means in addition to the main ring, the backup ring is also in use. In fact, the backup ring is used in order to maintain operation of the Token Ring without the stations in the troubled area. Figure 8–3 shows how the main ring is wrapped into the backup ring when there is a broken cable in the main ring.

Although each MSAU dedicates eight ports for the purpose of computer connections, the number of computers in the Token Ring network is not limited to eight. A single ring, as shown in Figure 8–1, can be extended to have up to 33 MSAU to support between 72 computers using UTP cable and 260 computers using STP cables. The speed of a Token Ring network is usually 4 Mbps using either a 100-meter UTP cable or a 200-meter STP cable between MSAUs, or 16 Mbps using a 75-meter UTP cable or a 100-meter STP cable between MSAUs. Figure 8–4 shows how a 16-Mbps Token Ring network can be extended by connecting the ring-out port of one MSAU to the ring-in port of another MSAU.

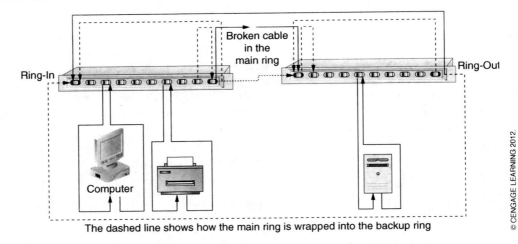

The dashed line shows how the main ring is wrapped into the backup ring

Figure 8–3: The main ring wrapped into the backup ring

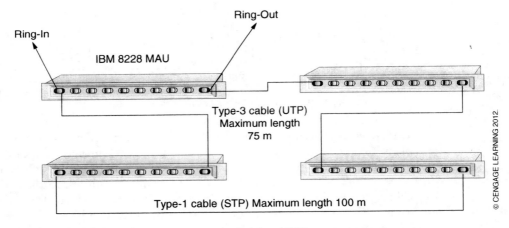

Figure 8–4: Extension of a Token Ring network using four MSAUs

8.3 INSERTION OF A NEW COMPUTER

To insert a new station (computer) into a Token Ring network, the following steps must be carefully taken:

1. Lobe testing (lobe_ media test). This is done by the NIC to verify the lobe media check. To perform this test, the new station is wrapped at the MSAU (looping the transmission line to the receiver line) to find out if the inserted station can transmit and receive signals without error.

2. Activate (open) relay. The NIC sends a 5-volt DC signal, which is called *phantom power*, on the transmission line to activate the relay at the MSAU port in order to insert a new station.

3. Becoming a member of the ring (address verification). After insertion of the new station into the ring, the inserted station will send a media access control (MAC) frame with its own MAC address in the destination address field of the Token Ring frame. The inserted station can be a new member of the ring if its address is unique in the ring. The return of the transmitted MAC frame to the inserted station will verify the uniqueness of its address.

4. Ring polling participation (neighbor notification). After the inserted station becomes a new member of the ring, it must participate in the ring polling process every 7 seconds. Through the ring polling process, the new station will find out who is participating in the ring and will learn the address of its nearest active upstream neighbor (NAUN).

5. Request initialization. In the final step, the new station must send the request initialization MAC frame to the ring parameter server (RPS). The RPS will respond with the initialization ring frame to provide the needed parameters such as a local ring number. The new station can be inserted into the ring with default parameters if the RPS does not respond.

8.4 CHARACTERISTICS OF THE TOKEN RING NETWORK

Unlike all networks that use the CSMA/CD access method, such as Ethernet, the Token Ring network is a deterministic type of network, which means it is possible to determine the maximum required time that a station must wait before it is able to transmit its frame. The Token Ring network also has a mechanism that allows stations to grab the token from the ring only if they have a priority equal to or greater than the priority value that is contained in the token. This mechanism is called the *prioritizing process*.

The Token Ring network is also capable of detecting and troubleshooting certain network faults according to an algorithm called *beaconing*. As was discussed earlier, there is an active monitor in each ring that sends a beacon frame every 7 seconds. The beacon frame is passed from one station to another. If there is a faulty station in the ring, it will not be able to respond to its upstream neighbor. In this situation, the upstream neighbor waits for 7 seconds. If it does not receive any response, it will notify the network of the lack of contact. It will then send a message that includes its address and the address of the neighbor station. Based on this information, the ring network will recognize the faulty station and attempt to diagnose and repair the problem.

A Token Ring usually uses type-1, type-3, or regular UTP cable. The type-1 cable is two 22 AWG (American Wire Gage) solid core pairs of STP cables with braided shields. Media Interface Connectors (MICs), which are the IBM type-A connectors, are used to connect type-1 cables to the MSAU. These connectors are neither male nor female. The type-3 cable is four 22 or 24 AWG UTP cables. Type-3 cables are voice-graded cables and, as we discussed earlier, UTP can only be used for transmission of data with speeds of 4 Mbps. The 8-pin 8P8C and 4-pin RJ-11 connectors are used for type-3 cables. To reduce line noise, a media filter is used between the Token Ring NIC and the 8P8C or RJ-11 connectors.

The Token Ring network works well with optical fiber cables because the Token Ring provides a high-speed unidirectional data flow.

Repeaters are used to extend the distance between MSAUs up to 1200 or 2400 feet apart, if type-3 or type-1 cables, respectively, are being used.

The Token Ring network uses the differential Manchester encoding method.

8.5 THE TOKEN RING FRAMES FORMAT

The Token Ring network has three types of frames: abort, token, and data/command frames. The IEEE 802.5 standard defines the Token Ring data frame format.

Abort Frame

The abort frame is a 2-byte frame which consists of the start delimiter (SD) and end delimiter (ED) fields only, and is used to abort (terminate) transmission signals from a station.

The SD field has a special bit pattern that indicates the start of the frame. The bit pattern of the SD field from the most to the least significant bits is: JK0JK000. Since the Token Ring network uses the differential Manchester encoding method, the non data J and K bits are selected to be differential Manchester violation bits (no transition at the middle of the J and K bits) in order to be detected by the hardware. The K bit is a steady high bit and the J bit is a steady low bit.

The ED field indicates the end of the frame, and its bit pattern from the most to the least significant bits is: JK1JK1IE. The I-bit is the intermediate bit, which indicates that a frame is part of the multiframe transmission. The E-bit is an error bit which will indicate the existence of an error in the frame. Figure 8–5 shows the abort frame format.

SD	ED
1B: JK0Jk000	1B:JK1JK1IE

© CENGAGE LEARNING 2012.

Figure 8–5: Abort frame © CENGAGE LEARNING 2012.

Token Frame

The token frame is a 3-byte frame which consists of the SD, access control (AC), and ED fields. The bit patterns of the SD and ED fields have already been shown. The AC bit pattern from the most to the least significant bits is: PPPTMRRR.

The P-bits are the priority bits of the token. The network administrators will assign a priority level (from the lowest level, 000, to the highest level, 111) to each station. For a station to seize the token from the ring, that station must have a priority level equal to or greater than the priority level of the token.

The T-bit is the token bit which indicates whether the frame is the token frame (T = 0) or the data/information frame (T = 1).

The M-bit is the monitor bit, which is used by the active monitor (AM) station to stop continuous circulation of a frame in the ring. When a station wants to send its frame, it sets the M-bit to 0, and when the frame passes from the active monitor, the AM sets the M-bit to 1. If the frame does not arrive at its destination and passes again from the AM, the AM station will remove the frame from the ring, purge it, and issue a new token.

The R-bits are the reserved bits that allow a station to reserve the priority of the next token that is going to be released by substitution of its priority bits into the reserved bits. The substitution can be done only if the station priority bits are greater than the existing reserved bits. Figure 8–6 shows the token frame format.

SD	AC	ED
1B: JK0Jk000	1B: PPPTMRRR	1B:JK1JK1IE

Figure 8–6: The token frame format

Data/Command Frame

In addition to the information that is transmitted from one station to another one, the data frame, as shown in Figure 8–7, contains: address fields, access control, frame control, error control, and the frame status.

SD	AC	FC	DA	SA	IF	FCS	ED	FS
1B	1B	1B	6B	6B	Includes LLC and MAC frames Size: up to 4500B	4B	1B	1B

Figure 8–7: The data/command frame

The bit patterns and functions of the SD, AC, and ED frames were discussed in the abort and token frame sections.

The frame control field (FC) is an 8-bit field with a bit pattern from the most to the least significant of: FFRRZZZZ (IBM frame) or FFZZZZZZ (IEEE 802.5 frame). FF bits indicate the frame type (FF = 00 means MAC frame, and the Z-bits, or control bits, indicate the type of MAC control frame; FF = 01 means logic link control (LLC) frame and the Z-bits will be ignored). The characteristics of the Z-bits are as follow:

ZZZZ = 0000 means Normal Buffer

ZZZZ = 0001 means Express Buffer

ZZZZ = 0010 means Beacon

ZZZZ = 0011 means Claim Token

ZZZZ = 0100 means Ring Purge

ZZZZ = 0101 means Active Monitor

ZZZZ = 0110 means Standby Monitor

The destination address field (DA) is a 48-bit field that indicates the physical address of the destination (recipient). The DA field in the Token Ring is similar to the DA field of the Ethernet. The most significant bit (the 48th bit) indicates whether the recipient station is an individual or a broadcast. The 47th bit indicates whether the recipient is universal or local.

The source address field (SA) is a 48-bit field that indicates the physical address of the source (transmitter). The SA field is very similar to the DA field except that the most significant bit (the 48th bit) is always set to 0 (IEEE 802.5 frame), which means the transmission always takes place from an individual source. The 48th bit in the IBM Token Ring frame is the routing bit which lets bridges either pass the token frame (bit 1) or ignore it (bit 0).

The data or information field (IF) contains the protocol data unit (PDU) and its size is variable (up to 4500 B), which depends on the speed of the ring. The FC field indicates the type of frame (MAC or LLC) and the IF field delivers the frame either to the MAC layer (IEEE 802.5) or to the LLC layer (IEEE 802.2).

The frame check sequence (FCS) is a 32-bit field that is used to check errors in the FC, DA, SA, and IF fields. The FCS field uses the CRC-32 error checking method.

The frame status (FS) is an 8-bit field that is used to let the source know whether or not its transmitted frame has been received at the destination and if the destination has copied it. The bit pattern of the FS field is shown in Figure 8–8. The A-bit indicates if any of the stations in the ring have recognized the address and the C-bit indicates if the frame was copied in the destination buffer. R-bits are the reserved bits. The frame status according to the A-bit and the C-bit is as follows:

AC = 00 No station has recognized the address and the frame was not copied

AC = 01 Not a valid combination

AC = 10 Destination has recognized the address but did not copy the frame

AC = 11 Destination has recognized the address and did copy the frame

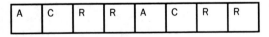

Figure 8–8: The bit pattern of the FS field © CENGAGE LEARNING 2012.

8.6 TOKEN BUS NETWORK

A Token Bus network is a broadband network and is defined by IEEE as the 802.4 standard. The broadband nature of the Token Bus network allows transmission over several channels at the same time. Transmission over several channels may cause traffic and collision. To manage traffic, the Token Bus standard defines four priority levels for traffic: 0, 2, 4, and 6, where level 6 is the highest level. The MAC layer assigns the priority level.

The operation of the Token Bus is similar to the Token Ring network but stations are numbered and physically connected to a long trunk coaxial cable in a star topology format. The stations are also logically connected to each other in a ring format. Therefore, token and data frames are passed sequentially from one station to another and follow a logical (virtual) ring. The logical ring may or may not include all stations sequentially. That means stations are connected to the bus sequentially but the token and data frame may circulate in a predetermined pattern. The last station in the ring will pass the token to the first station. Like the Token Ring network, each station in the Token Bus network must know the address of its neighbors. Figure 8–9 shows the typical Token Bus network.

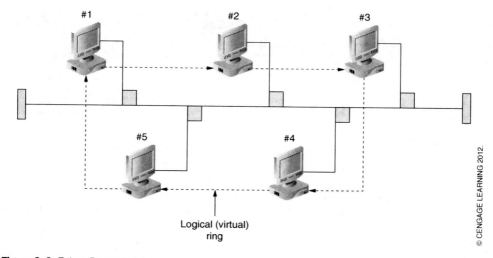

Figure 8–9: Token Bus network

In the Token Bus network, tokens pass from higher to lower addresses. When a station sizes the token, it has a limited time to transmit its frames. The number of transmitted frames depends on the size of each frame.

The Token Bus network is suitable for a manufacturing plant where stations may be located far from each other. General Motors has developed its Manufacturing Automation Protocol (MAP) based on the Token Bus network.

Token Bus Frame Format

The Token Bus frame format is similar to the Token Ring frame format with respect to the SD, FC, DA, SA, IF, FCS, and ED fields. The size of the IF and the error-checking method of the FC, however, are different than in the Token Ring network. The frame format of the Token Bus starts with the preamble field and is followed by SD, FC, DA, SA, IF, FCS, and ED. Figure 8–10 shows the Token Bus frame format.

Preamble	SD	FC	DA	SA	IF	FCS	ED
1B	1B	1B	2B or 6B	2B or 6B	0-8182B	4B	1B

Figure 8–10: Token Bus frame format

The preamble field is used to synchronize the receiver clock. The SD and ED fields indicate the start and end of the frame. They are analog encoding signals (non-data) which allow them to be distinguished from the content of the data frames.

The FC field indicates whether the frame contains data or control information. The control information of the FC is shown in Table 8–1.

FC field bit pattern	Name	Meaning
00000000	Claim_token	Claim token during ring initialization
00000001	Solicit_successor_1	Allow stations to enter the ring
00000010	Solicit_successor_2	Allow stations to enter the ring
00000011	Who_follows	Recover from lost token
00000100	Resolve_contention	Used when multiple stations want to enter the ring
00001000	Token	Pass the token
00001100	Set_successor	Allow stations to leave the ring

Table 8–1: The FC control information

The DA and SA fields contain a 2-byte or 6-byte destination or source address. The IF field has a variable size, which depends on the size of the address field and can exceed up to 8182 bytes of data. The FCS field is a 32-bit field that is used to check error using the checksum error-checking method.

SUMMARY

The Token Ring network was developed by IBM for its System Network Architecture (SAN) network and almost a decade later, it became the IEEE 802.5 standard. The Token Ring network consists of several stations (up to eight) that are connected to a multistation access unit (MSAU). To extend the number of stations in the ring, two or more MSAU units can be connected to each other by their ring-in and ring-out ports. The stations in the Token Ring network are physically connected to a hub (MSAU) in a star topology layout and are logically connected to each other in a ring topology layout.

One station in a ring acts as an active monitor and all other stations are the standby monitors. The active monitor is responsible for maintaining and managing the ring, detecting lost tokens, sending the beacon frame, and removing any continuously circulating frames.

The Token Ring network consists of two rings: a main ring and a backup ring. The main ring will be wrapped into the backup ring if there is trouble in the network, such as a broken cable.

Unlike Ethernet, the Token Ring network is a deterministic network. The insertion of a new station into the Token Ring network is relatively easier than for the Ethernet network.

The Token Ring frame consists of the following fields: start delimiter (SD), access control (AC), frame control (FC), destination address (DA), source address (SA), information field (IF), frame check sequence (FCS), end delimiter (ED), and frame status (FS). The IF may contain the MAC or LLC frames and can accommodate up to 4500 bytes.

The Token Ring uses CRC-32 for error checking and the differential Manchester encoding method.

The Token Bus is a broadband network in which stations are physically connected to each other in a bus topology layout and a virtual ring topology. The Token Bus network is a suitable system for manufacturing plants. The Token Bus is the IEEE 802.4 standard and its frame format is similar to but not exactly the same as the Token Ring network. The IF in the Token Bus network can accommodate up to 8182 bytes.

Review Questions

Questions

1. What types of physical and logical topology are used in the Token Ring network to connect stations to each other?

2. How many active monitors are in a Token Ring network?

3. What is the function of the active monitor?

4. What are the functions of the standby monitors?

5. How many stations can be connected to a single IBM 8228 MSAU?

6. How can two or more MSAUs be connected to each other?

7. How many rings are in a single Token Ring network?

8. What is the purpose of the backup ring?

9. What is the meaning of the term "purge the frame?"

10. What are the common speeds of the Token Ring network?

11. What are the common cable types of the Token Ring network?

12. What error-checking method is used in the Token Ring network?

13. What encoding method is used in the Token Ring network?

14. What are the priority bits and how do they help stations to size the token from the ring?

15. What is the purpose of the M-bit?

16. What is the purpose of the T-bit?

17. What are the purposes of the SD and ED fields?

18. How can a Token Ring network determine if the information frame contains the MAC frame or the LLC frame?

19. How will the Token Ring know if the destination is universal or local?

20. What is the indication of ZZZZ = 0010 in the frame control field?

21. What is the indication of AC 10 in the frame status field?

22. Is the Token Bus a baseband or broadband network?

23. What is the function of the preamble in the Token Bus frame?

24. What error checking method is used in the Token Bus network?

25. What is the meaning of the frame control field if its bit pattern is: 00000011?

9

Data Link Layer Protocols

Objectives

After completing this chapter, students should be able to:

- Describe data link layer (DLL) protocols.
- Discuss the flow and error control protocols.
- Explain the different types of Multiple Access protocols.

As was discussed in Chapters 6 and 7, IEEE has divided the data link layer into two sublayers: logical link control (LLC) and media access control (MAC). Flow and error controls are among the main responsibilities of the LLC sublayer and accessing shared media is the responsibility of the MAC sublayer. Many different protocols have been developed for the LLC and MAC sublayers. In this chapter these protocols will be discussed in detail.

9.1   FLOW CONTROL AND ERROR CONTROL PROTOCOLS

In a reliable communication network data must be transferred from a sender (transmitter) to a receiver successfully and efficiently. Transmission of data is successful if the sender and receiver possess the same data rate. In other words, if the data rate of the sender is greater than the data rate of the receiver, the receiver will not be able to copy the transmitted data in its buffer (memory storage) and, therefore, the transmitted data will be lost and need to be retransmitted. Retransmission may also be necessary if: (1) the receiver buffer size is not large enough to receive all of the transmitted data; (2) the data becomes corrupt when it reaches the receiver; or (3) the receiver's acknowledgment message does not reach the transmitter. Retransmission of data means inefficient use of the communication system. In the case of data lost at the receiver side, the receiver must inform the sender to stop sending new data and retransmit the lost data. The process of detecting lost data (error detection) and requesting retransmission of specific data is called *automatic repeat request* (ARQ). *Data link control* (DLC) is a term that is often used to define the tasks of flow control and error control.

The ARQ is a mechanism that informs the sender about the status of the transmitted data at the receiver's side by sending either a positive acknowledgment, ACK (the reception of data without an error), or a negative acknowledgment, NACK (the reception of data with an error). There are several different types of ARQ protocols that have been developed for the data link layer control in order to maintain proper data flow and error recognition. "Stop and Wait" is the first and simplest ARQ protocol. Continuous ARQ protocols that are also used in data communication and networking include Go-Back-N, Selective Repeat, and Polling.

9.2   STOP AND WAIT ARQ PROTOCOL

In this method, when the sender's data link layer receives packets of data from the network layer, it will format them into a frame, keep a copy of it, transmit it, and then wait for an acknowledgment message from the receiver. The receiver will send either a positive acknowledgment (ACK) if it has received the frame without errors, or a negative acknowledgment (NACK) if it has not received the transmitted frame (lost frame) or has received the frame with an error. The advantages of the Stop and Wait protocol are its simple implementation and congestion-free environment. Both half-duplex and full-duplex support the Stop and Wait protocol.

To maintain continuous data transmission, the sender will reset a timer and wait for a response from the receiver before the timer expires. If the receiver's response is positive (ACK), the sender will send the next frame. If the receiver's response is negative (NACK) or the receiver does not respond at all, the sender will retransmit the frame. There are four different scenarios in the Stop and Wait protocol:

1. The receiver receives a frame (frame number n) without error and sends a positive acknowledgment (ACK) to the sender. In this scenario, the sender then sends the next frame (frame number $n + 1$). Figure 9–1 shows the Stop and Wait protocol with no error.

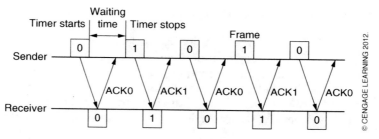

Figure 9–1: The process of the Stop and Wait protocol with no error

2. The receiver receives the transmitted frame with an error. In this scenario, the receiver sends a negative acknowledgment (NACK) and the sender then retransmits the frame (frame number n). Figure 9–2 shows the process of the Stop and Wait protocol with an error.

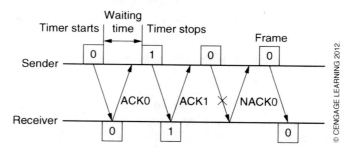

Figure 9–2: The process of the Stop and Wait protocol with an error

3. The transmitted frame is lost during transmission and the receiver has no knowledge about the transmitted frame. In this scenario, the sender will not get a response from the receiver and after the timer expires, the sender will retransmit the frame (frame number n). Figure 9–3 shows the process of the Stop and Wait protocol when the transmitted frame is lost during transmission and the timer expires.

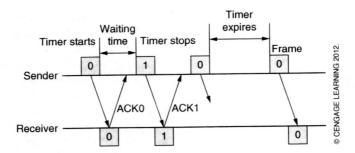

Figure 9–3: The process of the Stop and Wait protocol when the transmitted frame is lost

4. The receiver sends its acknowledgment (ACK or NACK) but the acknowledgment does not reach the sender before the timer expires, it comes with an error, or it is lost during transmission. In this scenario, the sender has to retransmit the frame (frame number n). Figure 9–4 shows the process of the Stop and Wait protocol when the acknowledgment frame is lost during transmission and the timer expires.

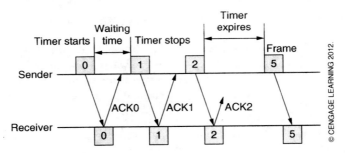

Figure 9–4: The process of the Stop and Wait protocol when the acknowledgment frame is lost

The above scenarios show that there are only two frames of interest in the Stop and Wait ARQ, the frame n and the frame $n + 1$, with only one frame outstanding at a time. That means assigning sequence numbers to frames is as simple as assigning sequence numbers of 0 and 1. In other words, Module-2 (Mod-2) arithmetic is used to number the sequence of frames in the Stop and Wait ARQ.

9.3 GO-BACK-N ARQ PROTOCOL

Go-Back-N is a continuous ARQ protocol where the sender sends frames in a specific time period or according to its window size and the receiver accepts the arriving frames, copies them in its buffer, and sends acknowledgments according to its window sizes. If one of the transmitted frames is lost during transmission or arrives at the receiver with an error, then the receiver will be faced with an out-of-sequence frame. At this time, the receiver will send a negative acknowledgment message (the "Rej" frame), request the sender to resend the corrupted or lost frame, and discard all frames that it received subsequently until it receives a frame with the expected sequence number.

Upon receipt of a Rej frame from the receiver, the sender will "go back" to the sequence frame starting with the corrupted or lost frame and retransmit the requested frame and all successive frames. This process is done by a technique called *sliding window*.

In the sliding window technique, both sender and receiver slide over their window and perform their task. The window size of the sender is $2^m - 1$ and the receiver window size is 1. For example, for $m = 3$, the sender has a window size of up to seven frames. The arrival packet from the network layer will be divided into a seven-frame window. The sender keeps sending its frames without waiting for acknowledgment from the receiver and slides over the rest of the frames that are to be transmitted. The receiver slides over only one frame and sends its ACK but if the arrival frame is out of sequence, the receiver either will be silent (if a timer is used) or send a NACK. Upon receiving a NACK or no acknowledgment if the timer expires, the sender will "go back" and retransmit all frames starting from the corrupted frame.

The Go-Back-N protocol uses a *pipelining* technique to perform its task. Pipelining is a technique in which a new task will be started before the old task is finished.

The Go-Back-N protocol also uses a technique that is called *piggybacking* to improve the efficiency of the bandwidth. In piggybacking, no special bandwidth is used for acknowledgment frames, which means the data and acknowledgment frames are transmitted together and not in a separate channel.

Figure 9–5 shows the process of the Go-Back-N ARQ protocol where the sender's window size is seven.

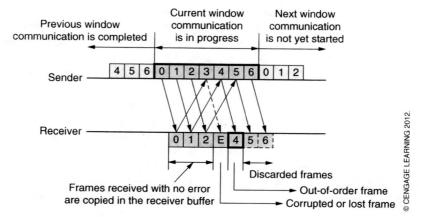

Figure 9–5: The process of the Go-Back-N ARQ protocol

As discussed in the data link layer specifications, the HDLC protocol of the data link layer contains a control field. The control field contains the information frame (I-frame), the supervisory frame (S-frame), and the unnumbered frame (U-frame). The I-frame contains data, the sequence number N(S), and the acknowledgment number N(R). The S-frame contains only the acknowledgment number N(R). The Rej is a case where the SS bits of the acknowledgment number in the I-frame and S-frame are 01.

9.4 SELECTIVE REPEAT ARQ PROTOCOL

The Selective Repeat ARQ protocol was developed to provide reliable and efficient communication. In this protocol, the receiver does not discard the uncorrupted, out-of-sequence frames. Instead, the receiver copies them into its buffer and the transmitter resends only the corrupted or lost frame. As a result, the channel's bandwidth is used more efficiently.

The window size for the sender and receiver is the same and is equal to 2^{m-1}, which is much smaller than the sender's window size in the Go-Back-N protocol.

The procedure for the selective repeat technique for a window size of seven is as follows:

1. The sender sends all seven frames and keeps a copy of all unacknowledged frames in its buffer.
2. The receiver checks the sequence number of the arrival frames. It will accept them if they have not already been received and store them in its buffer.
3. The receiver forwards its stored frames to the network layer only if there is no corrupted frame or out-of-sequence frames.
4. If frame number three is lost or corrupted during transmission:
   a. The receiver notices this when it receives frames 0, 1, 2, and 4. The receiver then sends a selective retransmission request (the "Srej" is a case where the SS bits of the acknowledgment number in the I-frame and S-frame are 11) to the sender for frame number 3 and continues to accept the out-of-sequence frames 4, 5, and 6 and copy them in its buffer.
   b. The sender retransmits frame number 3 when it receives the Srej message from the receiver and then continues to transmit its successive frames.
   c. The receiver accepts frame number 3, puts all of the seven frames in sequence and forwards them to the network layer.

The Selective Repeat protocol uses the pipelining technique to perform its task and the piggybacking technique to improve the efficiency of the bandwidth used.

The Selective Reject protocol may rely on a timer in its sender if the receiver does not respond or its response is lost during transmission. Again, if a sender does not receive any acknowledgment from the receiver when the timer expires, it will assume the receiver did not receive a particular frame and will retransmit it. Figure 9–6 shows the process of Selective Repeat ARQ protocol.

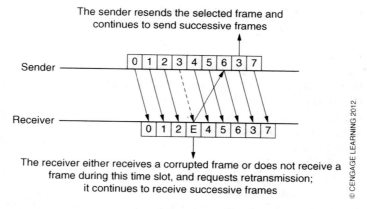

Figure 9–6: The process of the Selective Repeat ARQ protocol

9.5 ARQ FEEDBACK POLLING

Polling is a protocol that is used in primary–secondary or wireless networks. In the polling method, if the receiver does not receive a block of frames, or if the acknowledgment is not received by the sender, ARQ feedback polling is initiated. ARQ polling provides feedback to the receiver from the sender to ask whether or not the block of frames has been received. The sender then takes appropriate action based on the receiver's response. If the receiver's response is negative, the sender resends the blocks of frames. If the receiver does not respond again, then when the timer expires, the second ARQ feedback polling will be initiated. For example, if the sender sends a block of three frames (0–2) and the receiver does not return an acknowledgment before the timer expires, feedback polling will be initiated. If the receiver sends a negative acknowledgment, then the sender will resend all three frames. If the receiver sends a positive acknowledgment for the second run of transmission, the sender will continue to send its other frames. Otherwise the second feedback polling process takes place. Figure 9–7 shows the process of ARQ feedback polling.

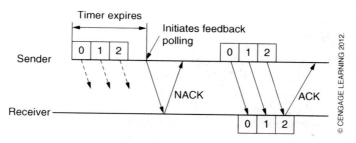

Figure 9–7: The process of ARQ feedback polling

9.6 MULTIPLE ACCESS PROTOCOLS

Communications between stations are either unicast or multicast. In unicast communication, a dedicated link or channel is established between two stations, whereas in multicast communication the channel is common to all stations. In other words, in multicast communication, multiple stations can access a common or shared channel at the same time. Different sets of rules (protocols) have been developed to avoid interferences or collisions when more than one station transmits messages or data frames over a shared channel simultaneously. These are called Multiple Access (MA) protocols. The MAC sublayer of the data link layer controls these protocols.

The MA protocols enhance the performance and robustness of the communication in a network and provide necessary information to a station that wants to transmit its data frame about the status of all the other stations in the network. The MA protocols are divided into three major categories:

1. Random Access Protocols
2. Polling or Control-Access Protocols
3. Channel Partitioning Protocols

Each of these is also divided into different subcategories.

Random Access Protocols

No prior conditions have been set among stations in the random access protocols. Therefore, each station has an equal right to transmit its frames at any time. This freedom of frame transmissions comes with a consequence, which is the possibility of collision between frames when they are transmitted at the same time. The random access protocol is a set of rules that allows stations in the network to detect and avoid traffic or collision and determines what happens in the event of traffic or a collision to make the channel ready to resume transmission. Several types of random access protocols have been developed over the years. The following methods are discussed: Pure-ALOHA, Slotted-ALOHA, Carrier Sense Multiple Access (CSMA), CSMA Collision Detect (CSMA/CD), and CSMA Collision Avoidance (CSMA/CA).

Pure-ALOHA

ALOHA was the first random access protocol, developed in early 1970 at the University of Hawaii. The first version of ALOHA is commonly called *Pure-ALOHA* (P-ALOHA) and its improved version is called *Slotted-ALOHA* (S-ALOHA).

In P-ALOHA, each station is able to transmit information regardless of other stations that also may send information. In this situation, the transmitted frame of information may or may not reach the destination. If the frame of information reaches the destination, the communication is a success and the transmitter will continue to send its other frames. But if the frame of information does not reach the destination (because of collision with another frame), the transmitter must retransmit the frame. To visualize how a collision of frames of information may happen during a transmission period, assume all stations in the network share the same data rate and the same frame size L. The time that it takes a frame to pass through a point in the channel is t. If the frame transmission time for stations A, B, and C are t_1, t_2, t_3, respectively, then to avoid collision between frames of stations A, B, and C, the difference between frame transmission times must be greater than t. Figure 9–8 shows conditions under which frames of three stations may or may not collide with each other in the P-ALOHA protocol.

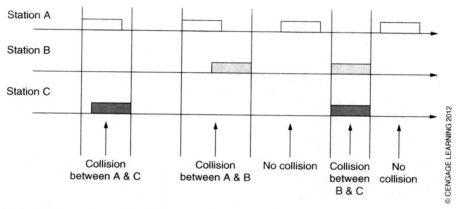

Figure 9–8: Collision and no-collision cases in the P-ALOHA protocol

The chance of collision in P-ALOHA is high and the maximum throughput value is about 18 percent. Therefore, this is not a suitable protocol for a high traffic network.

Slotted-ALOHA

Slotted-ALOHA (S-ALOHA) is the improved version of P-ALOHA. In this random access method, time is divided into slots of time T. If a station wants to send a frame, it must first gain access to the media and then transmit its frame at the beginning of an available time slot. If it misses the beginning of the available time slot, it has to wait for the next time slot. By dividing time into slots, the chance of collision is decreased compared with P-ALOHA. There is still a possibility of collision if two stations send frames at the beginning of the same time slot. Clock synchronization is needed in the S-ALOHA protocol because slots must be synchronized at the stations. Figure 9–9 shows how frames from different stations may collide in the S-ALOHA protocol.

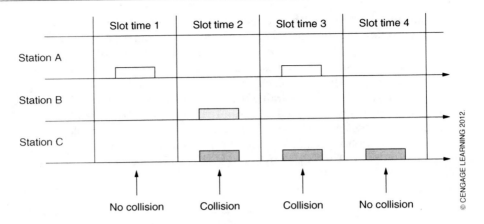

Figure 9–9: Collision and no-collision cases in the S-ALOHA protocol

Although slots may become vacant in the S-ALHOA protocol, dividing time into slots results in a higher maximum throughput, about 37 percent.

Carrier Sense Multiple Access

The Carrier Sense Multiple Access (CSMA) protocol was developed to improve the performance of frame transmissions based on collision reduction in the channel. To reduce collision in the channel, the transmitting station first senses the channel (checks the channel to determine whether it is busy or free for transmission) to find out if it is currently being used or if there is any collision before transmitting its frames. The process of sensing the channel in order to transmit frames if the channel is idle or busy is called *persistent*. Consequently, the CSMA protocol is divided into persistent and nonpersistent. Depending on the probability of a station sending its frame when the channel is idle, the persistent protocol is divided into two categories of 1-persistent and p-persistent.

In the nonpersistent protocol, if a station senses an idle channel, it transmits its frame immediately. If the channel is busy, the station waits a random amount of time and repeats the procedure of sensing the channel. Figure 9–10 shows the procedure. The random waiting time reduces the chance of collision but wastes channel idle time if the wait time is long.

In the 1-persistent protocol, when a station senses an idle channel, it transmits its frame with probability-1 (immediately). If the channel is busy, the station senses again with no waiting period. In this case, the probability of collision is high because other stations may also sense the idle channel and transmit their frames. The maximum throughput is about 55 percent, which is higher than the maximum throughput in ALOHA. The procedure of the 1-persistent protocol is shown in Figure 9–11.

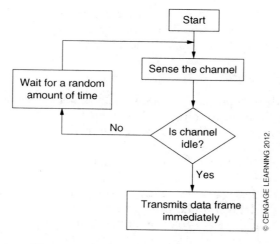

Figure 9–10: The procedure of the nonpersistent protocol

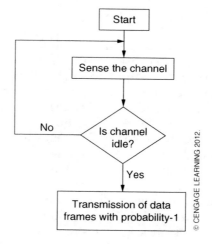

Figure 9–11: The procedure of the 1-persistent protocol

In the p-persistent protocol, the channel is divided into series of slot times. The size of the slot time is greater than the maximum propagation delay of one packet. When a station wants to transmit its frame it senses the channel. If the channel is idle, it transmits the frame with a predetermined probability P or waits one time slot for probability $1 - P$ before resensing the channel and repeating the procedure. The p-persistent protocol is a good compromise between the 1-persistent and nonpersistent protocols. The maximum throughput in the p-persistent protocol depends on the value of P. For example, the maximum throughputs for 0.5-persistent and 0.1-persistent are about 70% and 90%, respectively.

The CSMA with Collision Detection and CSMA with Collision Avoidance are two variations of the CSMA protocols. They were developed to either detect or avoid collision of frames in the media.

CSMA/CD Protocol

The CSMA/CD protocol is based on detecting collisions and terminating the transmissions in order to improve the channel capacity. CSMA/CD employs both persistent and nonpersistent transmission methods. The procedure of the CSMA/CD protocol is as follows:

1. A station first senses the channel. If the channel is idle, it transmits its frame.
2. If the channel is busy, it will continue to sense the channel until the channel is idle and then transmits its frame.
3. If there is a collision in the channel, it will stop transmission and send a jam signal to the network, wait for a random amount of time and repeat the procedure.

CSMA/CA Protocol

The CSMA/CA protocol was developed for wireless networks because it is not possible in a wireless network to sense the channel while transmitting frames. The procedure of the CSMA/CA protocol is as follows:

1. Set a counter that determines the number of times stations in the network sense and resense the channel.
2. If the channel is busy, a station waits for a predetermined time period, the inter-frame-space (IFS), and senses the channel again.
3. If the channel is still busy, it will try again. If the channel is idle, it waits for an additional time equal to a random number of time slots and transmits its frames.
4. The station waits for a back-off time to receive an acknowledgment.
5. If the acknowledgment is positive, the transmission is completed.
6. If the acknowledgment is negative, it will decrement the value on the counter and repeat the above procedure.

Polling Access Protocols

Polling access protocols are methods by which a station is invited by another station to access the media, or reserves or requests access. There are three types of polling access protocols: polling, reservation, and token passing.

Polling

Polling is an access protocol based on the master–slave or primary–secondary types of communication. In this protocol, the primary station is the master of the network and controls the network. The procedure for sending data frames is as follows:

1. The primary station can send its frames at any time and the channel becomes busy.
2. If the primary station does not have a frame for transmission, the channel is idle and primary. In this case, the primary polls the secondary stations one by one sequentially to determine if they have frames to send.
3. If the first secondary station has a frame to send, it responds with a positive acknowledgment (ACK) and sends its frames.
4. If it does not have a frame to send, it responds with a negative acknowledgment (NACK) and the primary polls the next secondary station and Step 3 is repeated.

The advantage of the polling protocol is in collision prevention by controlling the media so that only one station at a time sends frames. The disadvantages are in the polling overhead, latency, and failure of the entire network when the primary station fails.

Reservation

The reservation protocol is based on dividing time into slots and reserving slot time before transmitting data.

The interval of a time slot must be at least equal to the end-to-end propagation time in order to prevent collision between frames in two consecutive time slots. The procedure for sending data frames in this protocol is as follows:

1. Each station must reserve a slot time for transmission of its data frame.
2. All stations can see the reservation frame status, and each time slot can be reserved by only one station.

Figure 9–12 shows the process of reservation and transmission in the reservation protocol.

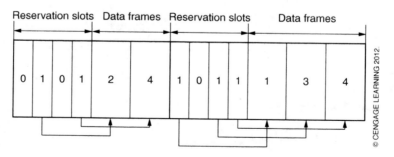

Figure 9–12: The process of reservation protocol

Token Passing

The token passing protocol is based on the circulation of a token in a logical ring and seizing the frame from the ring for the purpose of data frame transmission.

The active or primary station manages rights to seize the frame and the frame circulation. The procedure for this protocol is discussed in detail in Chapter 8.

Channel Partitioning Protocols

The channel partitioning protocols are based on channel sharing, which is done by dividing the channel into a series of time, frequency, or code slots where each specific slot is assigned exclusively to a single station. The channel partitioning protocols are divided into three categories: the Time-Division-Multiple-Access (TDMA) protocol, the Frequency-Division-Multiple-Access (FDMA) protocol, and the Code-Division-Multiple-Access (CDMA) protocol.

Time-Division-Multiple-Access

The Time-Division-Multiple-Access (TDMA) protocol partitions the bandwidth of the channel into several fixed time slots and each slot is assigned to only one particular station. The size of each time slot should be equal to the packet propagation time. Each station should know the start time of its own slot and send its frame during this specific time period. Synchronization between stations is required so that each station can successfully recognize the start of its slot and send its data frame. If a station does not have a frame to send, its slot will be empty in that specific round of data transmission. Unlike the TDM, which operates in the physical layer, the TDMA operates in the data link layer of the OSI model. Figure 9–13 shows the TDMA protocol for six stations in two consecutive transmission time periods.

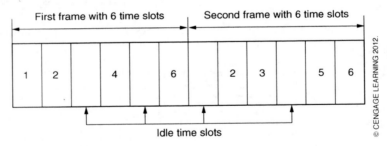

Figure 9–13: The TDMA format for a six-station network

Frequency-Division-Multiple-Access

In the Frequency-Division-Multiple-Access (FDMA) protocol, like in the TDMA protocol, the bandwidth is partitioned but using frequency band instead of time. Each frequency band is assigned to one particular station. The procedure in the FDMA protocol is very similar to that of TDMA. To prevent interference between stations, a band guard is inserted between each pair of frequency bands.

Code-Division-Multiple-Access

The Code-Division-Multiple-Access (CDMA) protocol is used mostly in wireless networks and is based on a code-assigning mechanism. A distinct code is assigned to each station that allows it to know what portion of the channel is available for use and for how long, in order to prevent collision in the channel. In the CDMA protocol, all stations can transmit their data frames simultaneously with minimum interference if their codes are orthogonal, which means the product of any two station codes is zero. Encoding and decoding of the station codes takes place at the transmitter and receiver, respectively. The Walsh table is commonly used to create the station codes. The Walsh table is based on a one-dimensional unit vector $[-1]$ or $[+1]$, and on the dimensional expansion of the vector based on the number of stations in the network. The following example demonstrates the creation of codes based on the Walsh table and by considering the unit vector of $[-1]$.

Example 9.1: (a) Use the Walsh table to develop codes for a network system with four stations. (b) Show the codes are orthogonal. (c) Show the encoding and decoding process. (d) Show the receiver will receive the exact bits through the CDMA multiple access method. Assume the four stations that share the channel have the following data for transmission during each 1-bit interval: bit 1, none, bit 0, and bit 1, respectively.

Solution: (a) Follow the Walsh rule to develop a four-dimensional vector in which each column represents a distinguishable code for one of the four stations in the network.

Start with the unit vector of $w = [-1]$ and the follow the Walsh rule to develop a two-dimensional and then a four-dimensional vector space.

$$W_1 = (-1)$$

$$W_2 = \begin{pmatrix} w_1 & w_1 \\ w_1 & /w_1 \end{pmatrix} = \begin{pmatrix} -1 & -1 \\ -1 & 1 \end{pmatrix} \text{ where } (/w_1) \text{ is the complement of } (w_1)$$

$$W_4 = \begin{pmatrix} w_2 & w_2 \\ w_2 & /w_2 \end{pmatrix} = \begin{pmatrix} -1 & -1 & -1 & -1 \\ -1 & 1 & -1 & 1 \\ -1 & -1 & 1 & 1 \\ -1 & 1 & 1 & -1 \end{pmatrix}$$

Each column represents a unique code. Assign columns from left to right to station A through D, respectively.

Code for station A = $[-1, -1, -1, -1]$ Code for station B = $[-1, 1, -1, 1]$

Code for station C = $[-1, -1, 1, 1]$ Code for station D = $[-1, 1, 1, -1]$

(b) Two codes are orthogonal if their inner product is zero. Show the codes for station A and B are orthogonal and follow the solution to show other codes are also orthogonal.

$$(\text{Code A}) (\text{Code B}) = \begin{pmatrix} -1 \\ -1 \\ -1 \\ -1 \end{pmatrix} (-1 \ 1 \ -1 \ 1) = (+1 - 1 + 1 - 1) = 0$$

(c) To encode the data of each station using its code, multiply the data value of each station by its corresponding code.

$$\text{Station A: } (1) [-1, -1, -1, -1] = [-1, -1, -1, -1]$$
where the data value of station A is one $(+1)$

$$\text{Station B: } (0) [-1, 1, -1, 1] = [0, 0, 0, 0]$$
where the data value of station B is none (0)

$$\text{Station C: } (-1) [-1, -1, 1, 1] = [1, 1, -1, -1]$$
where the data value of station C is zero (-1)

$$\text{Station D: } (1) [-1, 1, 1, -1] = [-1, 1, 1, -1]$$
where the data value of station D is one $(+1)$

(d) Add the leftmost bit of all four codes together to find the first bit in the transmission frame: $(-1 + 0 + 1 - 1) = (-1)$. Repeat this adding procedure for all other bits from left to right to find the other three bits in the transmission frame: $(-1 + 0 + 1 + 1) = (+1)$, $(-1 + 0 - 1 + 1) = (-1)$, $(-1 + 0 - 1 - 1) = (-3)$. Consequently, the transmission frame contains $(-1, +1, -1, -3)$.

At the receiver end, multiply the bits in the transmission frame (starting from the leftmost bit to the rightmost bit) by the codes of stations A, B, C, and D, respectively. Add the result and divide by 4 to find the data bit value of each station. The final result should match with the data bit of each station at the transmitter side.

$$(-1, +1, -1, -3) (\text{Code of station A}) = (-1, +1, -1, -3) [-1, -1, -1, -1]$$
$$= [1 - 1 + 1 + 3] = [4]$$

$[4]/4 = +1$ which means the data bit of station A is bit 1.

$$(-1, +1, -1, -3) (\text{Code of station B}) = (-1, +1, -1, -3) [-1, 1, -1, 1]$$
$$= [+1 + 1 + 1 - 3] = [0]$$

$[0]/4 = 0$, which means station B does not have a data bit.

$$(-1, +1, -1, -3) \text{ (Code of station C)} = (-1, +1, -1, -3) [-1, -1, +1, +1]$$
$$= [+1 - 1 - 1 - 3] = [-4]$$

$[-4]/4 = -1$, which means the data bit of station C is bit 0.

$$(-1, +1, -1, -3) \text{ (Code of station D)} = (-1, +1, -1, -3) [-1, 1, 1, -1]$$
$$= [1 + 1 - 1 + 3] = [4]$$

$[4]/4 = +1$, which means the data bit of station D is bit 1.

SUMMARY

To facilitate the performance and efficiency of a network system that operates under the OSI model, the IEEE set a standard by dividing the data link layer into two sublayers: logical link control (LLC) and media access control (MAC). The flow and error control data frames are among the responsibilities of the LLC sublayer. The MAC sublayer manages how multiple stations access to the link or channel.

Several protocols have been designed and developed to control data flow and errors in transmission. Stop and Wait is a protocol in which the sender sends its data frame and waits for acknowledgment (either positive, ACK, or negative, NACK) from the receiver and repeats this procedure until all its data frames have been transmitted. In the Go-Back-N protocol, the sender does not wait for an acknowledgment for each frame but if it receives a NACK for any of its frames, it will resend all frames starting from the corrupted frame. In the Selective Repeat, the sender resends only the frame that was corrupted during transmission. ARQ feedback polling is used in wireless networks and is based on the master–slave method of communication. In this protocol, the sender initiates a polling mechanism for feedback from the receiver. If it receives a negative acknowledgment it resends the block of data frames.

Three major protocols also have been designed and developed to facilitate and control access of multiple stations in a shared channel. Random access, polling or control, and partitioning are the main protocols, and each has several different methods. The P-ALHOHA, S-ALOHA, CSMA, CSMA/CD, and CSMA/CA are among the random access protocols. In the P-ALOHA, all stations transmit frames, and therefore there is a high risk of collision between frames. In S-ALOHA, stations can send their frames in their own slots. In all CSMA protocols, a station must sense the channel before it sends its frame. In the CSMA/CD protocol, the collision is detected and a notify message is transmitted to all stations. In the CSMA/CA protocol, the collision is avoided.

Reservation and polling are among the control-access protocols. In the reservation protocol, stations must reserve a slot before sending their data frames. In the token passing protocol, a token will be circulated (passing from one station to the next station) and stations must seize the token first and then transmit their data frames.

The channel partitioning protocols are based on channel sharing. In these protocols, the channel capacity is divided into series of time (TDMA), frequency (FDMA), or code (CDMA) slots where each specific slot is assigned exclusively to a single station.

Review Questions

Questions

1. What is the function of flow control?

2. What is the function of error control?

3. What are the three common flow and error control mechanisms?

4. Define *continuous ARQ* and name the different types.

5. Why do we need a timer in a flow control mechanism?

6. What is the difference between Go-Back-N and Selective Repeat ARQs?

7. What are the advantages and disadvantages of the continuous ARQ?

8. Define piggybacking.

9. What is pipelining?

10. What are the multiple access protocols? Describe each category and their types.

11. What is the main difference between the P-ALOHA and S-ALOHA protocols?

12. Under what condition may a collision happen in the S-ALOHA protocol?

13. How does the reservation protocol operate?

14. How can a station send its data frame in a token passing protocol?

15. What is the persistence strategy and how do the two persistence strategies differ?

16. What is the difference between 1-persistent and p-persistent?

17. How does CSMA function?

18. Is there any limit for a station to sense the channel in CSMA?

19. What is the main difference between CSMA/CD and CSMA/CA?

20. What would happen if a station senses a collision in the channel if CSMA/CD has been used?

21. Define the channel partitioning multiple access protocols.

22. In what types of communication are the channel partitioning protocols used?

Continues on next page

23. What are the similarities and differences between TDMA and FDMA?

24. How are codes generated for the CDMA protocol?

25. In what communication system is CDMA mostly used?

Problems

1. Draw the process of the Stop and Wait protocol when a station sends five data frames and the second and fourth arrive at the receiver with errors.

2. Repeat Problem 1 for Go-Back-N ARQ in which only the third data frame arrives with an error.

3. Repeat Problem 2 for the Selective Repeat ARQ.

4. A network has eight stations and uses the reservation protocol for accessing the shared channel. Show graphically the reservation slots and the transmitting frames for two consecutive time intervals when the first, second, fourth, sixth, and seventh stations send their data frames in the first time interval and the third, fifth, seventh, and eighth stations send their data frames in the second time interval.

5. Show graphically the format of the TDMA with eight slots of time where the second, third, fourth, sixth, and seventh stations send their data frames in the first time interval and the first, third, fifth, seventh, and eighth stations send their data frames in the second time interval. Which slot times are idle in each time interval?

6. Start with the unit vector of $w = [+1]$ and do the following:

   a) Use the Walsh table to develop codes for a network system with four stations. Show the codes are orthogonal.
   b) Show the encoding and decoding process.
   c) Show the receiver will receive the exact bits through the CDMA multiple access method.

   Assume the four stations that share the channel have the following data for transmission during a 1-bit interval: none, bit 1, bit 0, and bit 1, respectively.

10

Network Devices

Objectives

After completing this chapter, students should be able to:

- *Discuss the characteristics and operation of various network devices.*

- *Describe the architectures of a network using routers, switches, hubs, repeaters, and bridges.*

- *Discuss the operation of gateways, firewalls, network cards, and interface cards.*

A network may consist of hundreds of stations (nodes, devices, or computers). Data frames are broadcast from one station to another. The territory of this network is called the *broadcast domain*. The number of collisions between data frames increases as the number of stations increase in a broadcast domain with a single collision domain. To reduce the chance of collision in a broadcast domain, the collision domain is divided into several subcollision domains with fewer stations. Dividing a network in to multiple subcollisions is also called segmentation of the network. Figure 10–1 shows a broadcast domain with two subcollision domains.

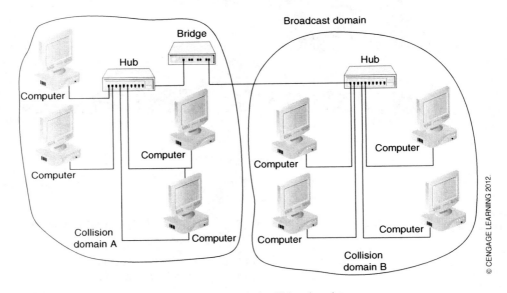

Figure 10–1: Dividing a broadcast domain into two subcollision domains

A number of network devices have been designed and manufactured to connect the subcollision domains in a network. The function of each device is unique and is used for a specific task and in a specific layer of the OSI model. These include:

1. Layer one devices: repeaters and hubs
2. Layer two devices: bridges and switches
3. Layer three devices: routers

Other network devices include firewalls, gateways, and network interface cards (NICs). These different network devices are reviewed in this chapter.

10.1 REPEATERS

A repeater is the same as a regenerator, which accepts an attenuated (degraded) signal due to long cable lengths and/or the number of stations within the network, reshapes the attenuated signal through amplification and retiming, and finally retransmits it to other stations. Repeaters do not filter the signal. Therefore, if a signal contains noise or other disturbing signals, these will be amplified as the signal undergoes amplification. The span of the network can be extended by regenerating the traveling signal in a network through repeaters, but the extension should be limited in the preassigned subcollision domain. A network may require multiple repeaters to extend its span, because a signal can travel only for a limited distance before becoming attenuated. However, the number of repeaters in a network is limited due to timing and other issues.

Repeaters do not control traffic, collision, or error in the network; they deal only with signals and therefore operate at the physical layer. Figure 10–2 shows the operation of a repeater in a network.

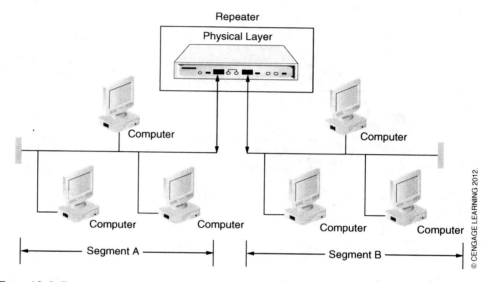

Figure 10–2: The operation of a repeater in a network

10.2 HUBS

A hub is a network device that is able to connect multiple stations together to facilitate design and building of a small network such as a home or small office network. A hub is a repeater with multiple ports that receives data frames from one station and transfers them out to all stations within that particular network via its ports. Hubs do not filter the incoming or outgoing frames and they are not able to reroute or redirect the incoming data frames. A hub is a simple and inexpensive device for expanding a network without boosting its performance. Connecting stations in a network without hubs is difficult and time-consuming, and most importantly creates noise and error in the network. Hubs are not the network device to be used for network segmentation. They increase the useable distance of the network, but do not enhance the network performance since they are unable to do packet processing. Hubs are unable to connect different network architectures and do not reduce network traffic. The number of hubs in a network must be limited in order to prevent noise and to keep the chance of collision low.

Hubs are the heart of the star network topology. They use the half-duplex transmission mode, which means they cannot send and receive information at the same time. Hubs share the bandwidth and operate in a single collision domain. Hubs operate in the physical layer of the OSI model and, therefore, the data transmission in the hub is in bits. Hubs do not have a fixed speed; speed can vary from 10 to 100 Mbps. Figure 10–3 shows the operation of hubs in a network.

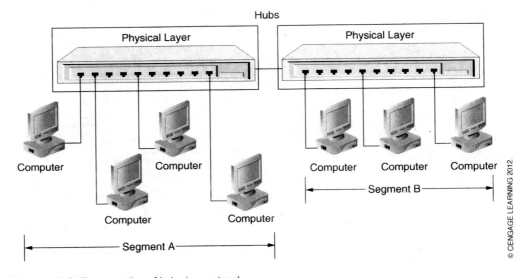

Figure 10–3: The operation of hubs in a network

Over the years, the function of hubs has expanded to perform some other functions of a network system. Today, three different types of hubs are available in the market: passive, active, and intelligent hubs.

1. Passive hubs are the original type of hubs which are used only to connect multiple stations together.
2. Active hubs, which are also called multiport repeaters, are able to amplify weak incoming signals and convert them to stronger signals for retransmission to all other stations. Active hubs are also able to filter out noise from the signals.
3. An intelligent hub not only has the advantages of the passive and active hubs, but also is able to monitor and manage traffic in the network in order to provide a better and more efficient network.

10.3 BRIDGES

A bridge is a network device that is used to connect two self-governing network segments together to be able to exchange data. The connecting networks could be different types but they should use compatible media access control (MAC) addresses, media, and communication protocols. A long-span network may also be divided into several smaller segments and connected via bridges to reduce traffic and enhance the efficiency of the network.

A bridge is an active device. This means that the bridge examines the incoming information, which includes the source and destination addresses, and compares them with an address table. The address table is either given by the system manager or the bridge sends out a broadcast message to all stations within the local area network (LAN) and asks them to identify themselves. The bridge passes the information if the destination address belongs to other network segments, and ignores it if the address of the destination is in the same network segment. To pass the information to another network segment, the media must be in an idle position. If it is not, the bridge must store the information and wait until the medium is idle before it passes the information. To store the incoming information, the bridge should have a buffer.

Bridges operate at the data link layer (layer 2) of the OSI model, and for this reason they are also called *layer-2* switches. Figure 10–4 shows the operation of a bridge in a network.

A bridge is a hardware device but its operation is guided by software.

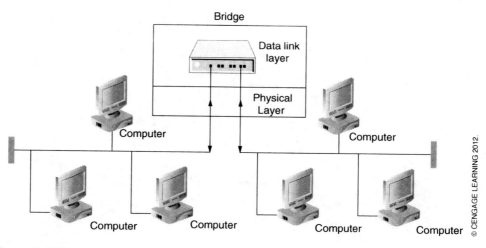

Figure 10–4: The operation of a bridge in a network

A source routing bridge is a specific type of bridge that is used in a token ring network. It is called the source routing bridge because the source identifies the route that a data frame must take to reach the destination. To find out which is a better route to take, the source sends out a discovery frame to the entire network. The discovery frame will identify all of the possible routes, record them, and provide them to the source.

10.4 SWITCHES

Switches are network devices that are used to connect multiple stations together within a network via their ports. Switches operate at the data link layer (layer 2) of the OSI model and are designed to enhance the performance of a LAN network by establishing a virtual circuit between source and destination without sharing links with other stations. Switches are also able to establish simultaneous transmission and reception of data frames between stations in the network. Switches have their own bandwidth and are full-duplex. They are able to reduce the workload on individual stations and decrease the number of packet collisions. Switches are more expensive than hubs and their function is more complex. Thus, troubleshooting their network connectivity is more difficult. Figure 10–5 shows the operation of a switch in a network.

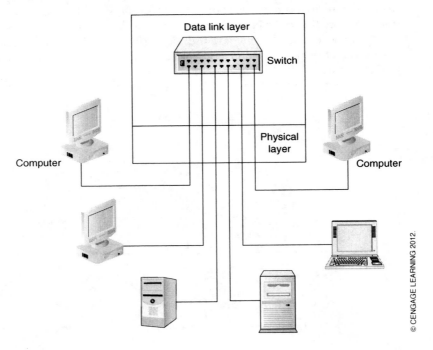

Figure 10–5: The operation of a switch in a network

The function of a layer-2 switch is to forward traffic to the destination host only, according to the CAM table. The CAM table stores information such as MAC addresses, switch ports, and virtual LAN (VLAN) parameters. When a layer-2 switch receives a frame, it searches for the destination MAC address in the CAM table and takes the following actions depending on whether or not the destination MAC address is recognized.

1. If the destination MAC address is recognized, the switch forwards the frame of data.
2. If the destination MAC address is recognized and is identical to the source MAC address, the switch filters out the frame.
3. If the destination address is not recognized, the switch transmits the frame to all connected ports except the originating port.

The switch forwards the frame to the recognized destination using one of the following methods:

1. The store-and-forward switching method, which enhances the performance of the network by reducing the packet-handling time (latency). In this method, the switch in the network receives data packets from the source station, drops the invalid packets, stores them in its buffer, and finally forwards the data packets to the destination. The forwarding decision is based on receiving the entire frame and checking its reliability.
2. In the cut-through method, the switch in the network receives data packets from the source station and forwards them to the destination immediately after it reads, examines, and stores only the destination address. The cut-through method does not carry out the error-checking process.
3. In the fragment-free method, the switch in the network reads and examines the data packets but does not carry out the error-checking process.
4. In the adaptive switching method, the switch in the network automatically switches back and forth between the other three methods.

Stations in a LAN network are connected to each other either by fully connected or switched-circuit methods. The following four different types of switched-circuit methods are used in a LAN network.

1. Circuit switching provides a dedicated link, which is also called *microsegmentation*, between two stations for a communication session. The microsegmentation helps switches to create collision domains so that only the NICs of the source and destination can contend for the link. When the communication session is over, the link is available for the next communication session. In other words, circuit switching is a session-based communication method that establishes a dedicated communication link between two stations within a network. Circuit switching is used in telephone communication.
2. Message switching accepts messages (information) from the source and stores them in its buffer and then forwards them to the destination. The requirement for a large buffer is a disadvantage of this type of switched-circuit method.
3. Virtual switching establishes different transmission paths for the packets of information before the communication takes place between source and destination.

4. Packet switching enables the source to divide information into several individual packets and label them with a sequence of numbers. The information packets travel along different paths from one switch to another and are reassembled at the destination according to their sequence of numbers. Each switch must have a small buffer to store incoming packets of information when the connecting path is busy. Use of packet switching in a network helps to maximize the bandwidth efficiency, minimize the transmission latency, and increase robustness of communication. Packet switching is used in many current communication protocols such as TCP/IP. Figure 10–6 shows a typical packet-switching method.

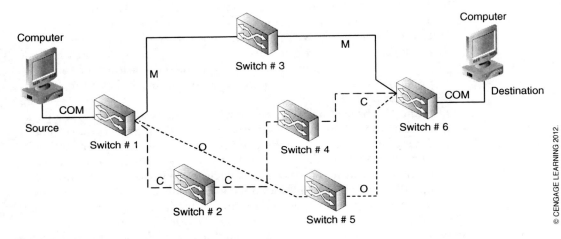

Figure 10–6: A typical packet-switching method

10.5 ROUTERS

A router is a network device that operates at the network layer (layer 3) of the OSI model and is able to forward data packets according to the routing table from one network to another. Data packets can be moved between LAN and LAN, LAN and WAN, or LAN and Internet service provider (ISP) network. In other words, a router is an intermediate (gateway) network device between an external network such as an Internet or broadband connection, and an internal network such as a home network. If two different networks (for example, Token Ring and Ethernet networks) must be connected to each other via a router, the router should be able to convert the Ethernet frame format to the Token Ring frame format and vice versa. The router functions as a central network device that helps to share an Internet connection, files between computers, and printers. Some routers have an embedded DSL modem, as well as firewall capabilities and advanced security features. Figure 10–7 shows the operation of a core router in a single network configuration.

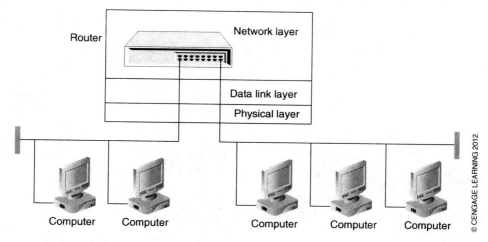

Figure 10–7: The operation of a core router

A router is called a static router if the routing table is configured by the network administrator, and dynamic if the routing table can be configured by the router itself. Most routers have built-in Dynamic Host Configuration Protocol (DHCP) servers that help a computer be configured automatically (i.e., assigning a correct IP address, recognizing their domain or server names) without getting any help from the network administrator. To determine the best route to forward data packets, the routers are equipped with headers and forwarding tables, and they use protocols such as an Internet Control Message Protocol (ICMP) to arrange the best route of communication between any two hosts.

A standard network interface card was used when routers were first designed. But later, when the demand for higher-speed data packet transmission grew, routers were designed with a forwarding engine embedded in the network interface. Configuration of a router is a complex task requiring routing software. Routers cannot be used in a network that does not provide the needed information for routing. Different types of routers include the following:

1. Core routers, which forward data packets between stations within a network.
2. Edge routers, which forward data packets between stations in two or more networks.
3. Bridge routers (brouters), which function as router and bridge in a network.
4. Virtual routers are backup routers used in a Virtual Router Redundancy Protocol (VRRP) environment. In the VRRP environment, multiple routers are connected to each other virtually to create a Virtual Private Network (VPN) in order to prevent data loss during transfer. One of these virtual routers controls the IP addresses and is called the *master router*,

and others are the backup routers. The master router is in charge of successful forwarding of the data packets. If a backup router in this environment fails to function within a certain period of time, the other backup routers will watch the master router for commands. If the master router fails, one of the backup routers will take over the function. A priority system is implemented in a VRRP environment between the backup routers.

5. Wireless routers are used in a wireless network. The wireless routers can be connected to the modem to provide Internet access to the network or they can come with an embedded modem to simplify the network connection and reduce the number of devices in a network. Figure 10–8 shows a small office network with a wired router and a wireless router.

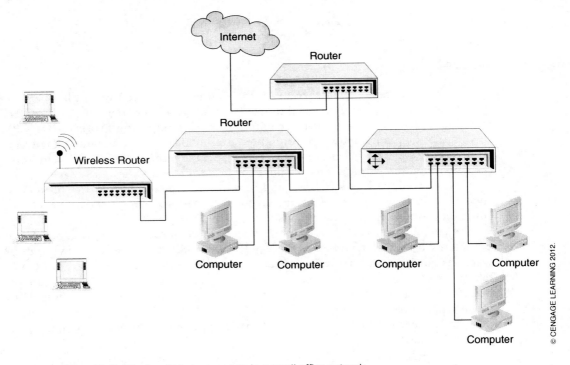

Figure 10–8: Wired and wireless routers in a small office network

10.6 GATEWAYS

A gateway is an entrance/interworking node that connects two networks with different base protocols together and operates in all seven layers of the OSI model. Gateways are implemented entirely via hardware, entirely via software, or via a mixture of software and hardware. Gateways, in association with routers and switches, help a computer in a network to forward the data packets in an actual route via the ISP to the Internet and vice versa. At the home network level, a modem or a broadband router can be used as a gateway to communicate over the World Wide Web by asking for an IP address from the ISP's nearby router. In an office network, the gateway may also function as a firewall and a proxy server. Figure 10–9 shows the operation of a gateway in the OSI model.

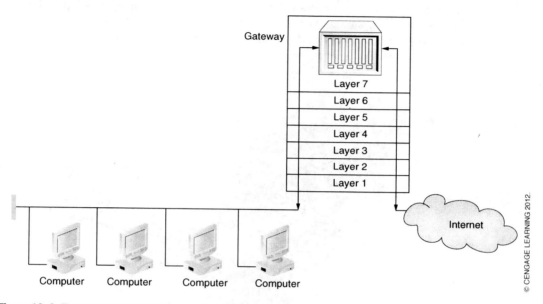

Figure 10–9: The operation of a gateway in the OSI model

10.7 FIREWALLS

A firewall is a security device or a program that protects a network by inspecting and filtering the incoming data packets or applications and either allowing them to enter the network or rejecting them. A network firewall is able to secure a network by preventing it from unauthorized access and blocking malware. It is also able to screen and limit access to external networks for the users within the network. Firewalls are implemented in hardware or software or a mix of software and hardware. Software firewalls may include antivirus, anti-popup, and antispam and they are usually used in home or small office networks. Hardware firewalls are mostly used in large networks. A firewall may be placed in the network proxy server or gateway.

10.8 NETWORK CARDS, ADAPTORS, AND NETWORK INTERFACE CARDS

A network card, network adapter, and network interface card (NIC) are the computer hardware devices that are used for different purposes in a network. The network card functions in the physical layer of the OSI model and provides security at this layer. A network adapter functions at the data link layer of the OSI model to provide data security for the computer. The NIC provides a hardware interface between a computer in the network and the network itself. The NIC functions in both the physical and data link layers. It accepts data from a computer in the network and converts it into the data frame and then to a signal able to be broadcast in the network media. NICs are designed to connect computer bus types such as USB or PCI to the network media. The communication between NICs and the network is established by either the NIC driver or firmware software. Figure 10–10 shows a typical NIC available in the market.

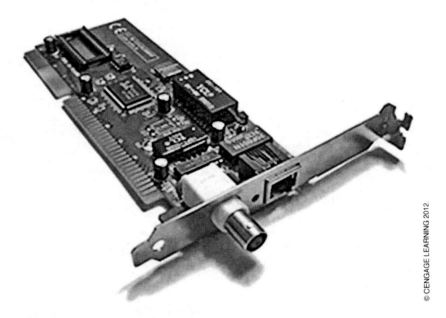

© CENGAGE LEARNING 2012.

Figure 10–10: A typical network interface card

SUMMARY

A variety of network devices are needed to design and build a network system. The network devices are designed to perform different tasks in a network to establish reliable and efficient transfer of data packets in a network or between two or more networks, or just to expand the network. Network devices function at different layers of the OSI model. More than one network device may function in the same layer of the OSI model, but perform different tasks. For example, repeaters and hubs both function in the physical layer (layer 1), but repeaters are responsible for regeneration of the transmission signal to be processable at the receiver, whereas hubs are mostly used to receive the signal and distribute it to other stations. Bridges and switches are functions in the data link layer (layer 2), and routers function in the network layer (layer 3).

Switches forward data packets using four different methods: store-and-forward, cut-through, fragment-free, and adaptive. The four different switches are circuit switching, message switching, virtual switching, and packet switching. Packet switching in a network helps to maximize the bandwidth efficiency, minimize the transmission latency, and increase robustness of communication. Routers are intermediate network devices between an external network and an internal network. There are five different types of routers: core, edge, bridge (brouter), virtual, and wireless.

The gateway helps users to access the Internet. The firewall is a security device that prevents unauthorized users from accessing the network or filters out unwanted networks. Network interface cards connect the computer bus type to the network media. Communication between the NIC and the network is established by either the NIC driver or firmware software.

Review Questions

Questions

1. What are the main network devices?
2. What is the function of a router?
3. What is the function of a hub?
4. In what layer of the OSI model does the repeater function?
5. In what layer of the OSI model does the bridge function?
6. In what layer of the OSI model does the NIC function?
7. What is the difference between a repeater and a hub?
8. Explain the function of the store-and-forward switch.
9. What are the most popular switches? Name them all.
10. What is the main difference between core and edge routers?
11. What is the function of a firewall?
12. What is the function of the NIC driver?
13. Is firmware software or hardware?
14. What is the function of the network adapter?
15. How can a user access the Internet?
16. How can a network administrator filter or block a specific Web site?
17. Explain how packet switching works.
18. What are the advantages of packet switching?
19. What is the function of the bridge router?
20. How can you divide the collision domain of a network?
21. In what layer of the OSI model does the switch function?
22. How does the cut-through switch work?
23. How does the fragment-free switch work?
24. In what layer of the OSI model does the router function?
25. Design a network with eight stations, two hubs, and one router. All stations need to be able to access the Internet.

11

The Transport Control Protocol (TCP) and Internet Protocol (IP)

Objectives

After completing this chapter, students should be able to:

Describe Transport Control Protocol (TCP).

Describe Internet Protocol (IP).

Discuss the operation of the network access layer and the protocols that operate in it.

Transport Control Protocol (TCP) and Internet Protocol (IP) are the standard protocols that were developed in the early 1980s for the Internet. The Internet actually evolved from the research of the Department of Defense's Advanced Research Projects Agency (ARPA) on network technology which was called ARPANET. The IP and TCP protocols have different responsibilities. The IP is responsible for the data packet movement from node to node based on the IP address. The IP address is a 4-byte host address that is connected to the Internet. Under the IP operation, data packets move from a host in a network around the world via internetworks. The TCP is responsible for safely and completely delivering the data packets from host to host. To holistically understand the TCP/IP protocol, it is necessary to know the Internet reference model.

11.1 THE INTERNET REFERENCE MODEL

The Internet reference model, like the OSI model, is a layer-based model and consists of the following four layers:

1. Network access layer
2. Internet or internetwork layer
3. Transport or host-to-host transport layer
4. Application layer

In comparison to the OSI model, the Internet reference model has combined the physical and data link layers into one single layer, the network access layer. The Internet and the transport layers correspond to the network and transport layers of the OSI model. And finally, the application layer combines the session, presentation, and application layers of the OSI model. IP operates in the Internet layer and TCP operates in the transport layer. Each of the Internet, transport, and application layers has different protocols with different functions.

HyperText Transfer Protocol (HTTP), File Transfer Protocol (FTP), Simple Mail Transfer Protocol (SMTP), and Telnet are the four best known protocols of the application layer. HTTP supports file (including text and graphics) transports between the client and server. FTP supports the end-to-end file transfer, and SMTP supports basic message delivery. Telnet lets the user implement terminal sessions with remote hosts. Other protocols that operate in the application layer include Simple Network Management Protocol (SNMP), Domain Name Service (DNS), and Trivial File Transfer Protocol (TFTP).

The stage (characteristic) of the traveling data packets changes as they pass from the application layer to the lower layers by the addition of a specific piece of information to the data packets called the *header*. This process is called *encapsulation*. The headers will be opened and read at each layer of the recipient terminal and they will be disregarded at their respective layers. At the application layer, the data packets are in their original stage (message) and they will be segmented at the transport layer. At the Internet layer, they are at the datagram stage and they reach their final stage at the network access layer, which is known as a *frame*. Table 11–1 shows a comparison of the Internet reference model and the OSI model.

OSI model	Internet reference model	Protocols	Stage of the data
Application layer	Application layer	HTTP, Telnet, SMTP, and FTP	Data packet (message)
Presentation layer			
Session layer			
Transport layer	Transport layer	TCP, UDP, and RTP	Segment
Network layer	Internet layer	IP, ICMP, ARP, and RARP	Datagram
Data link layer	Network access layer	Ethernet, Token Ring, ATM, FDDI, X.25, Frame Relay, RS-232, NIC, and PPP	Frame
Physical layer			

Table 11–1: Comparison of the Internet reference model with the OSI model

11.2 THE NETWORK ACCESS LAYER

The network access is the lowest layer of the Internet reference model and it is responsible for completely and safely delivering data from one computer to another via a physical medium. The responsibilities of the network access layer are as follows:

1. Accepting data from the Internet layer in the form of datagrams, and encapsulating datagrams into frames for transmission over a physical medium.
2. Mapping the IP address in the network to the physical address of the recipient computer, which has been burned into the network adapter card.
3. Learning the necessary information such as the physical address of the destination, the maximum frame size, and the data packet structure for proper and reliable data delivery.
4. Learning the specific details of the physical medium such as type of medium and the exchange of data to signal and vice versa.
5. Routing data via the connecting media and appending its routing information as a header to the transmitting frame.
6. Detecting errors.

The frame format of the network access layer consists of the IP datagram, the Media Access Control (MAC) frame header, and the frame check sequence. The information field is located in the IP datagram. Figure 11–1 shows the frame format of the network access layer.

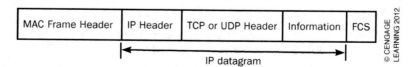

Figure 11–1: The network access layer frame format

Protocols that operate in the network access layer are Ethernet, Token Ring, Asynchronous Transfer Mode (ATM), FDDI, Frame Relay, RS232, and Point-to-Point Protocol (PPP). The Ethernet, Token Ring, and RS232 were covered in Chapters 1, 7, and 8. The other protocols are not in the scope of this book but brief explanations are presented here.

Asynchronous Transfer Mode

The Asynchronous Transfer Mode (ATM) is a high-speed packet (cell) switching transfer mode that operates at the data link layer of the OSI model. The information, which may include data, voice, or image, is arranged into cells with a fixed length of 53 bytes, including 48 bytes of data and 5 bytes of header information. The fixed cell size provides a fast switching and multiplexing function which leads to high-speed data transfer and also provides a self-routing mechanism. The ATM is a connection-oriented service that establishes a point-to-point connection via ATM switches before data transmission. The ATM protocol provides two types of interfaces, network-to network interface (NNI) or user-to-network interface (UNI), and consists of three layers which, from top to bottom, are:

1. The physical layer is responsible for cell delineation (describing the cell), cell verification (header error control), and placing and removing cells from the physical medium. This layer adds an idle cell if it is required. The physical layer supports different physical media and different media interface rates. The physical layer maps the cells into time-division multiplexed frames.

2. The ATM layer is responsible for performing multiplexing, switching, and monitoring of the quality of service (QoS). It is also responsible for cell generation and extraction, cell flow, and providing the cell transfer mechanism using Virtual Path and Virtual Circuit (VP/VC) routing.

3. The ATM adaptation layer (AAL) provides service to upper and lower layers and is responsible for cell segmentation (segmenting information to 48 bytes), reassembly, timing, and flow control. The AAL, is divided into two sublayers: the convergence and segmentation sublayer and the reassembly sublayer (CS and SAR). There are three types of AAL, which provide different services: type 1 or constant bit rate (CBR) service, type 2 or variable bit rate (VBR) service, and type 3 or connection-oriented data service.

Frame Relay

Frame Relay is a fast and efficient information packet transmission method which was originally designed for an Integrated Service Digital Network (ISDN) based on X.25. (X.25 is the original packet switching technique that uses a telephone line or ISDN connection as a physical link. It was developed before the development of the

OSI model.) In the Frame Relay protocol, data pass between local area networks or end points in a wide area network in HDLC packets called *frames*. Frame Relay is a packet switching protocol for digital communication which operates at the physical and data link layers of the OSI model. This means packet transmission takes place at the data link layer rather than at the network layer. Two different packet techniques, variable-length and statistical multiplexing packets, are used in the protocol. There is no acknowledgment from the receiver in the Frame Relay protocol, and if there is an error in the packet or there is network congestion, the transmitted packets will be dropped. This means error correction takes place at the receiver, not the transmitter.

Transmission of packets between stations in Frame Relay occurs via virtual circuits. The virtual circuits are frequently changing logical paths transparent to all stations. Frame Relay employs two different virtual circuits: Permanent Virtual Circuits (PVCs) for a dedicated point-to-point connection, and Switched Virtual Circuits (SVCs) for call-by-call communication.

Point-to-Point Protocol

Communication between a computer in a network and the Internet can be established by either a Point-to-Point Protocol (PPP), which is provided by the local Internet service provider (ISP), or via the network interface card. In the Internet reference model, PPP provides a connection service at the network access layer. PPP can carry out both synchronous and asynchronous communication and works with different types of transmission media such as twisted pair wire or optical fibers, and also wireless transmission. It is a full-duplex protocol and has error detection ability.

11.3 INTERNET OR INTERNETWORK LAYER

The internetwork layer accepts the segmented data packets from the network layer, encapsulates them in a datagram form, and directs them through internetworks. *Datagram* is the name given the encapsulated data packets at the Internet layer. A datagram consists of information, TCP or UDP header, and its own header. In short, the internetwork layer performs the following tasks: defining the datagram, defining the Internet addressing scheme, accepting data from the network access layer and transferring them to the transport layer, routing datagrams to remote hosts, fragmenting datagrams in order to route them in different networks, and reassembling datagrams.

The header of the internetwork layer contains the destination and source addresses for routing the packets of information through the network, as well as other information such as security. The internetwork layer does not perform error detection and correction. If the receiver of the datagram encounters an error, it simply drops it. Figure 11–2 shows the internetwork layer datagram format.

IP Header	TCP or UDP Header	Information

Figure 11–2: The internetwork layer datagram format © CENGAGE LEARNING 2012.

The protocols that operate at the internetwork layer are as follows:
1. Internet Protocol (IP)
2. Internet Control Message Protocol (ICMP)
3. Address Resolution Protocol (ARP)
4. Reserved Address Resolution Protocol (RARP)

Internet Protocol is a connectionless protocol, which is responsible for transferring the data packets from node to node using a system of logical host addresses known as the IP addresses. IP makes the routing decision for each data packet. IP is an unreliable protocol for error detection and correction and depends on other layers to perform this task. It also depends on other layers to establish a connection if there is a need for such a service.

Flow control, detection of the unreachable destination, redirecting routes, and checking the remote hosts are among the tasks of ICMP.

ARP is basically used to map the IP address to the MAC address, which is utilized by the data link sublayer of the network access layer. RARP reverses the ARP function by mapping the MAC address to the IP address.

The original IP, known as IP Version-4 (IPv4), has a 4-byte address that contains the network, host, or multicast group identification numbers (IDs) depending on what class of IP addresses are used. IPv4 has five different addressing modes or classes (A, B, C, D, and E). The content of the first byte in an IP address indicates the class of the IP address. The following are the characteristics of each class.

Class A is used in a unicast communication network with a large number of users. Its frame format starts with logic bit 0 and is followed by a 7-bit network ID and 24-bit host ID. An IP address is also class A if the content of the first byte is written in decimal numbers (0–127).

Class B is used in a medium-sized unicast communication network. Its frame format starts with a binary number of 10 and is followed by a 14-bit network ID and 16-bit host ID. An IP address is also class B if the content of the first byte is written in decimal numbers (128–191).

Class C is used in a unicast communication network with a small number of users. Its frame format starts with a binary number of 110 and is followed by a 21-bit network ID and 8-bit host ID. An IP address is also class C if the content of the first byte is written in decimal numbers (192–223).

Class D is used in the multicast communication network and its frame format starts with a binary number of 1110 followed by a 28-bit multicast group ID. An IP address is also class D if the content of the first byte is written in decimal numbers (224–239).

Class E is reserved for future use. Its frame format starts with a binary number of 1111 followed by a 28-bit multicast group ID. An IP address is also class E if the content of the first byte is written in decimal numbers (240–255).

Example 11.1: Determine the class type of the following IP addresses. X represents a 2-bit binary number.

Example	1st Byte	2nd Byte	3rd Byte	4th Byte
a	216	XXXX	XXXX	XXXX
b	00XXX	XXXX	XXXX	XXXX
c	1100XX	XXXX	XXXX	XXXX
d	248	XXXX	XXXX	XXXX
e	1110XX	XXXX	XXXX	XXXX
f	168	XXXX	XXXX	XXXX

© CENGAGE LEARNING 2012

Solution:

a. Class C, because the decimal number 216 is in the class C range (192–223).
b. Class A, because the first byte starts with 0.
c. Class C, because the first byte starts with 110.
d. Class E, because the decimal number 248 is in the class E range (239–255).
e. Class D, because the first byte starts with 1110.
f. Class B, because the decimal number 168 is in the class B range (128–191).

The IP addresses are used by the internetwork and higher layers to identify devices and to perform internetwork routing. The following are some of the important characteristics of IPv4. The address size for both source and destination is a 32-bit (4-byte) address.

1. Supports a 576-byte packet size with fragmentation.
2. Configured either manually or through Dynamic Host Configuration Protocol (DHCP).
3. IP security (IPSec) is optional.
4. Header contains a checksum, but does not indentify packet flow for QoS handling by routers.
5. The sending host and routers do the data packet fragmentation.
6. Different IP addresses are used for networks with a different number of users.
7. Sends traffic to all nodes on a subnet by using the broadcast addresses.

The source and destination address, fragmentation, routing, and flow are among other functions of IPv4 that are included in the header. The header total length is 32 bits. Figure 11–3 shows the header format of IPv4.

Version (4-bit)		Header Length (4-bit)	Type of Service (8-bit)	Total Length (16-bit)
Identifier (16-bit)		Flags (3-bit)	Fragment Offset (13-bit)	
Time-to-Live (8-bit)		Protocol (8-bit)	Checksum (16-bit)	
Source Address (32-bit)				
Destination Address (32-bit)				
Optional (24-bit) (if any)		Padding (8-bit) (if any)		
Data (variable)				

© CENGAGE LEARNING 2012.

Figure 11–3: The IPv4 header frame

The identifier is used for fragmentation. The time-to-live is used to make sure packets are forwarded (how long they exist to the network). The network must be decremented at each router. Packets with a time-to-live of zero (TTL = 0) are thrown away.

The first bit of the flag field is not used and the second bit is used to indicate whether or not the datagram can be fragmented. The third bit is used to indicate if there is a need for more datagram fragmentation. The checksum field is used to monitor the integrity of the header and to find out if it has been corrupted during transmission.

The disadvantages of IPv4 are the timitation of IP addresses, that the network configuration is done manually or through DHCP, and the security issues. To overcome these shortages the Internet Engineering Tasks Force (IETF) has developed Internet protocol version-6 (IPv6) using a 128-bit address, which provides more than 3×10^{23} IP addresses. The IPv6 header contains a flow label field which identifies packet flow for QoS handling by a router. Figure 11–4 shows the IPv6 header format.

Version (4-bit)	Traffic Class (8-bit)	Flow Label (20-bit)
Payload (16-bit)	Next Header (8-bit)	HOP Limit (8-bit)
Source Address (128-bit)		
Destination Address (128-bit)		

© CENGAGE LEARNING 2012.

Figure 11–4: The IPv6 header format

Under IPv6, a single data packet can be sent to multiple destinations with the same high QoS. The following are some of the important characteristics of IPv6:

1. Supports IPSec.
2. Header does not contain a checksum but does identify packet flow for QoS handling by routers.
3. The sending host fragments data packets; routers do not.

4. Supports guided configuration.
5. The address size for both source and destination is 128-bit (16-byte).
6. Supports a 1280-byte packet size without fragmentation.

11.4 HOST-TO-HOST TRANSPORT LAYER

The host-to-host transport layer, also called the transport layer, of the Internet Reference Model provides end-to-end reliable data delivery and communication service for nodes that demand extended bidirectional data exchange. The data delivery is based on the open and close commands. At the beginning of the open command, a host-to-host connection or virtual connection will be established between computers, and data transfer in the form of a stream of characters will begin. Data transfer ends at the time of a close command. Unlike the OSI model, there is no session layer in the Internet Reference Model. The responsibilities of the session layer are performed by the transport layer, and instead of "session" the words "socket" or "port" are used to describe the route of communications.

Transmission Control Protocol (TCP) and User Datagram Protocol (UDP) are the most common protocols used at the transport layer. TCP is used when data error correction and resubmission of the data are required and quarantined. TCP provides full-duplex and reliable connections and enables hosts to maintain multiple and simultaneous connections. Unlike TCP, the UDP is used when a reliable connection service is not necessary. This means that error correction is not an issue, and it is easier to retransmit data than to detect and correct it. This is the case when the amount of data transferred is not huge. The error control field takes a good portion of the data frame, which results in a high overhead issue.

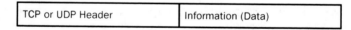

| TCP or UDP Header | Information (Data) |

Figure 11–5: The transport layer frame format © CENGAGE LEARNING 2012.

Real-Time Transport Protocol (RTP) is another protocol that operates at the transport layer. It provides end-to-end real time data transport of multimedia data (data, audio, and video) via unicast or multicast network services. RTP, along with Real-Time Transport Control Protocol (RTCP), provides data transportation to large multicast network services. RTCP monitors the services of RTP and provides feedback quality. Internet telephony is one of the most common applications of RTP. Figure 11–5 shows the host-to-host transport layer frame format.

11.5 APPLICATION LAYER

The application layer of the Internet Reference Model is responsible for standardizing the presentation of data and providing the services that user applications need to communicate over the network. This means that the application layer contains all application protocols that use host-to-host data transportation. Most of the applications in this layer are network service types of applications and usually are either file transfer types or connection services.

The most common protocols that operate at the application layer are:

1. Telnet, which helps users perform sessions with remote hosts.
2. Simple Mail Transfer Protocol (SMTP), which provides basic data delivery.
3. Hyper Text Transfer Protocol (HTTP), which provides low-overhead file (data and image) delivery.
4. File Transfer Protocol (FTP), which provides basic interactive file transfer between hosts.
5. Domain Name Service (DNS), which maps IP addresses to the names assigned to the network devices.
6. Simple Network Management Protocol (SNMP), which helps collect management information from the network devices.

SUMMARY

The Internet Reference Model is a four-layer model that provides standard protocols for the Internet. The four layers of this model are: the network access layer (layer 1), the internetwork layer (layer 2), the host-to-host transport layer (layer 3), and the application layer (layer 4). The network access layer corresponds to the physical and data link layer of the OSI model, the internetwork layer corresponds to the network layer, the host-to-host transport layer corresponds to the transport layer, and the application layer of the Internet Reference Model corresponds to the session, presentation and application layers of the OSI model. The transport layer of the Internet Reference Model also performs the responsibilities of the session layer of the OSI model.

Each layer is responsible for specific tasks and has a number of protocols to facilitate the performance of each layer. The network access layer is responsible for delivering data between computers completely and safely via a physical medium. The frame format of this layer consists of the MAC frame header, IP header, TCP or UDP header, information, and frame check sequence (FCS). The Ethernet, Token Ring, ATM, FDDI, Frame Relay, RS232, and PPP are among the protocols that operate at this layer.

The internetwork layer is responsible for defining the datagram, defining the Internet addressing scheme, accepting data from the network access layer and transferring them to the transport layer, routing datagrams to remote hosts, fragmenting datagrams in order to route them in different networks, and reassembling datagrams. IP, ICMP, ARP, and RARP are the common protocols that operate at this layer. IP is responsible for providing data packet delivery between two nodes in the network.

The host-to-host transport or transport layer is responsible for providing end-to-end reliable data delivery and communication service for nodes that demand an extended bidirectional data exchange. TCP, UDP, and RTP are the most common protocols that operate at this layer. The TCP is responsible for completely and safely providing data packet delivery from host to host.

The application layer is responsible for standardizing the presentation of data and provides the services that user applications need to communicate over the network. Telnet, SMTP, HTTP, FTP, and DNS are the most often used protocols that operate at this layer.

Review Questions

Questions

1. What are the differences between the OSI and the Internet Reference Model?

2. What are the most common protocols that operate at the internetwork layer?

3. What is the main function of IP?

4. What is a datagram?

5. What is encapsulation?

6. What are the most common protocols at the host-to-host transport layer?

7. What is the main function of the network access layer?

8. Does the network access layer perform the same tasks as the physical layer in the OSI model? Explain your answer.

9. Briefly explain the function of ATM.

10. How many layers exist in the ATM?

11. At what layer of the OSI model does the Frame Relay operate?

12. How many virtual circuits does the Frame Relay use?

13. What is the main function of the internetwork layer?

14. How many classes are used in IPv4?

15. How can you figure out which class is used in an IP address?

16. Which class is used for multicast network services?

17. What class is used if the IP address starts with a binary number of 10?

18. What class is used if the IP address starts with a binary number of 1110?

19. What class is used if the IP address starts with a decimal number of 202?

20. What class is used if the IP address starts with a decimal number of 248?

21. What is the total size in length of the IPv4 header?

22. What is the time-to-live?

23. What is the function of the flag field in the IPv4 header?

24. What are the main advantages of IPv6 over IPv4?

25. What is the main difference between TCP and UDP?

26. What is the function of RTP?

27. What is the main function of the application layer?

28. What is the function of HTTP?

29. What is the function of DNS?

30. What are the open and close commands?

12

Hands-On Activities for the Networking Part of the Course

Objectives

After completing this chapter, students should be able to:

▨ *Create and manage a new user account.*

▨ *Find the IP and the physical address of the network adapters.*

▨ *Create a new connection to the Internet.*

▨ *Allow other computers to share the Internet connection.*

▨ *Troubleshoot problems with the network connection.*

Determine the configuration of a wireless network connection, its setting, properties and specifications.

Connect to a local area network and configure a wireless connection, its setting, properties and specifications.

Add a client to the network in order to have access to the shared resources on the network.

Use ping to test a network connection.

Build and test a network using a daisy chain.

A well-equipped network laboratory is essential to understand the functionality of each network protocol and devices that are employed in that network. This type of laboratory is costly and not every college or university that offers a networking course would be able to have such a laboratory. Fortunately, it is possible to perform some network related experiments using only Microsoft Windows or a similar package. In this chapter, nine such laboratory experiments are presented to help students understand the topic of networking by conducting these experiments. In addition, a hands-on experiment that uses hardware is also presented at the end of this chapter.

All activities in this chapter use Windows 7, and most require you to have Administrator permissions to complete them.

12.1 HANDS-ON ACTIVITY 1: CREATING AND MANAGING A NEW USER ACCOUNT

1. Click on the Windows logo in the lower left corner of the screen and choose Control Panel from the Start menu.
2. In the Control Panel window, click the Add or remove user accounts link under the User Accounts and Family Safety heading.
3. Click the Create a new account link.
4. Enter a user name for the new account and select the Standard User radio button. Click on Create Account.
5. In the Manage Accounts window that opens, click the icon for the new user.
6. There are several options for managing the new account, such as: create a password, change the picture, set up Parental Controls, change the account type, and delete the account.
7. To create a password for the new account, click Create a password. A new window opens. Type your password in the box and confirm it. Click Create password to complete creating a new user account with a password (see Figure 12–1).

Figure 12–1: Creating a password for a new user account

12.2 HANDS-ON ACTIVITY 2: WINDOWS INTERNET PROTOCOL (IP) CONFIGURATION

1. Click on the Windows logo in the lower left corner of the screen.
2. Type **cmd** in the search box in the bottom left corner of the Start menu. A Command Prompt window will open on your desktop.
3. At the C:\ prompt, type **ipconfig/all**. Information regarding the Windows IP configuration, such as the physical address for the wireless local area network (LAN) adapter, Ethernet adapter, Ethernet adapter Local Area connection, and IP Version-4 (IPv4) will appear (see Figure 12–2).

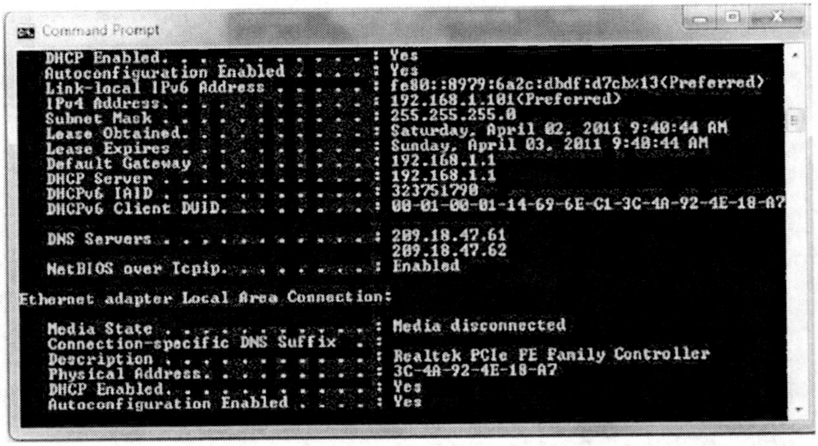

Figure 12–2: IP configuration

12.3 HANDS-ON ACTIVITY 3: VIEW YOUR CONNECTION TO THE INTERNET

1. Click on the Windows logo in the lower left corner of the screen and choose Control Panel from the Start menu.
2. In the Control Panel window, click on the Network and Internet heading.
3. Click the View network status and tasks link under the Network and Sharing Center heading (see Figure 12–3).
4. Spend some time investigating the different options in the Change your networking settings area of the Network and Sharing Center. (To ensure that you don't make any unintended changes, use the Cancel button to exit out of any new windows you open during your investigation.)

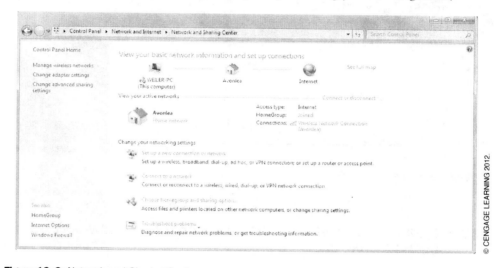

Figure 12–3: Network and Sharing Center

12.4 HANDS-ON ACTIVITY 4: SHARING THE INTERNET CONNECTION

1. Open the Network and Sharing Center as you did in Hands-On Activity 3.
2. Click the Network Connection link in the View your active networks area of the Network and Sharing Center window. (Depending upon how you are connected, this will say something like Wireless Network Connection, as shown in Figure 12–3.) This will open the Network Connection Status window, as shown in Figure 12–4. This window contains all general information about the network connection, including IP Version-4 (IPv4) or IP Version-6 (IPv6), media status, SSID, and signal quality.
3. Click the Properties button to open the Network Connection and Properties window. Select the Sharing tab and check the Alllow other network users to connect through this computer's Internet connection box. Click OK and close all open windows.

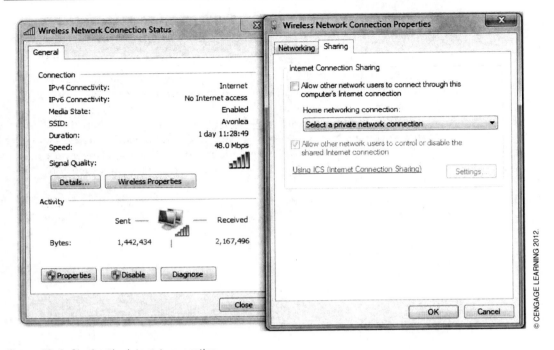

Figure 12–4: Sharing the Internet connection

12.5 HANDS–ON ACTIVITY 5: TROUBLESHOOTING A NETWORK CONNECTION

1. Open the Network and Sharing Center as you did in the two previous hands–on activities.
2. Click the Troubleshoot problems link in the Change your networking settings area of the window.
3. In the Troubleshoot problems—Network and Internet window that opens, select Incoming Connections.
4. In the Incoming Connections window, click the Advanced link. For the purposes of this activity, uncheck the Apply repairs automatically checkbox. Click Next.
5. In the What are you trying to do? window that opens, select the Find this computer on the network radio button, and click Next.
6. The computer will automatically start to troubleshoot the problem. It will take a few seconds to complete this process. Once the process is complete, Windows provides a Troubleshooting report (see Figure 12–5).

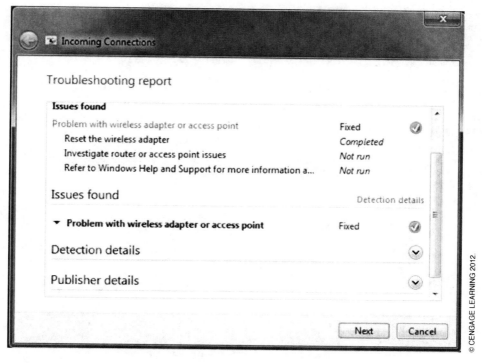

Figure 12–5: Troubleshooting network report

12.6 HANDS–ON ACTIVITY 6: WIRELESS CONFIGURATION

1. Open the Network and Sharing Center as before.
2. Select the Change adapter settings link shown on the left side of the Network and Sharing Center window.
3. The Network Connections window showing your available connection options opens. If you are using a wireless connection, select the Wireless Network Connection. (For the purpose of this exercise, you can either double-click on the connection name, or single-click and select View status of this connection from the menu bar.)
4. Click Properties in the Wireless Network Connection Status window (see Figure 12–4).
5. In the Networking tab of the Wireless Network Connection Properties window, click Configure.
6. Select the Advanced tab. This will display the list of your network connection properties, such as the type of network media (version of the IEEE 802.11 for wireless network media), and the receiver (download

rate) and transmitter (upload rate) buffers. See Figure 12–6 for an example of wireless configuration properties.

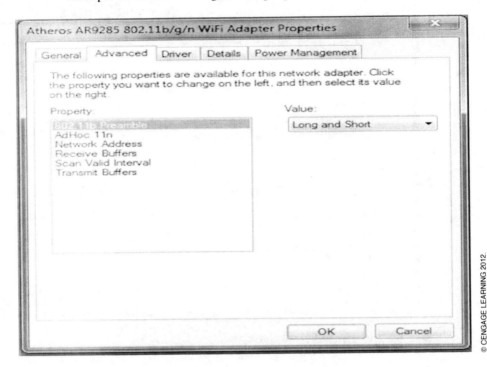

Figure 12–6: Wireless configuration

12.7 HANDS–ON ACTIVITY 7: LOCAL AREA NETWORK CONFIGURATION

1. To perform this hands-on activity, you will need to be plugged into a network with a network cable. Additionally, you will need to know your network address, which was displayed in the Command window in Hands-On Activity 2 (see Figure 12–2). (You may need to obtain your network address from your workplace network administrator).
2. Open the Network and Sharing Center as before.
3. Select the Change adapter settings link shown on the left side of the Network and Sharing Center window.
4. The Network Connections window showing your available connection options opens. Select the Local Area Connection. (For the purpose of this exercise, you can either double-click on the connection name, or single-click and select View status of this connection from the menu bar.)
5. Click Configure in the Networking tab of the Wireless Network Connection Properties window.

6. Select the Advanced tab. This will display the properties that can be set for a local area network (LAN). Scroll down the property list and click on Network Address. In the Value box, enter the IP address of your network. See Figure 12–7 for an example of local area network configuration properties.
7. You may repeat the above steps to check other properties of the LAN after you have connected to it.

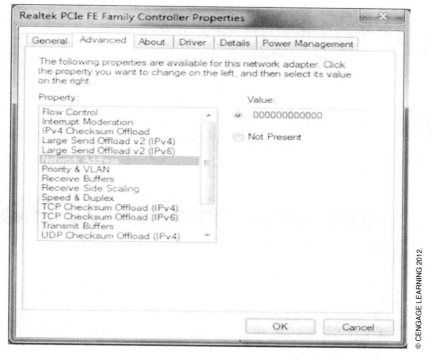

Figure 12–7: Local area network configuration

12.8 HANDS-ON ACTIVITY 8: ADD A CLIENT TO THE NETWORK

1. Open the Network and Sharing Center as before.
2. Click on the Network connection link to open the Wireless Network Connection Status window.
3. Click Properties to open the Wireless Network Connection Properties window.
4. Select Client for Microsoft Network from the list of connection items, and click Install...to open the Select Network Feature Type dialog box.
5. Select the Client option and click Add. A client provides access to computers and files on the network you are connected to. (See Figure 12–8a.)

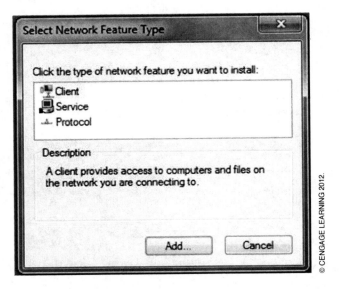

Figure 12–8a: Add a client to the network

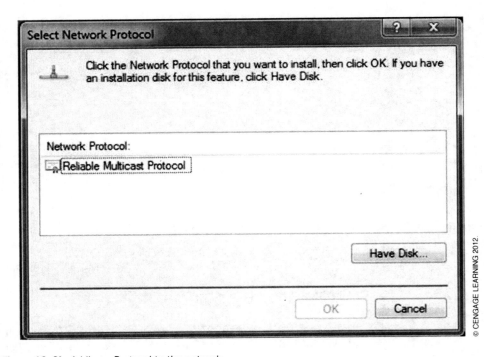

Figure 12–8b: Adding a Protocol to the network

6. Insert the Windows installation disk, and in the Select Network Client window, click Have Disk. The Windows 7 Installation Wizard will guide you through the appropriate steps to complete the process of adding a client.
7. You may also install Service or Protocol to the network by repeating the above steps, and selecting the appropriate option from the Select Network Feature Type dialog box. A Service provides additional features such as printer and file sharing. This can also be done in the Network and Sharing Center. A protocol is a language your computer uses to communicate with other computers and, in the case of Windows 7, it is a multicast protocol (see Figure 12–8b).

12.9 HANDS-ON ACTIVITY 9: USING PING TO TEST A NETWORK CONNECTION

1. Ping is a command used to determine if another computer is available on a network. This could include either your intranet or the wider Internet. You must know either the host name of the computer you wish to connect to or the computer's IP address. For this activity, let's use a common IP address: 192.168.1.1.
2. Click the Windows logo in the bottom left corner of the screen to open the Start menu.
3. Type **cmd** in the search box to open a Command Prompt window.
4. At the C:\ prompt, type **ping 192.168.1.1** and press Enter.
5. All ping statistics, including the sent and received packet size, packets lost, and the round-trip time will appear in the Command Prompt widow (see Figure 12–9).

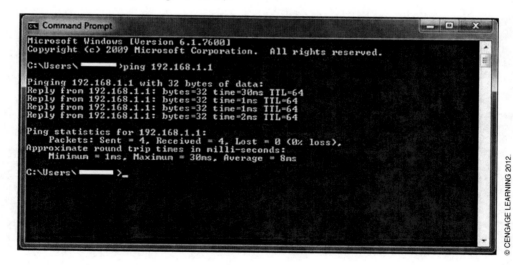

Figure 12–9: Using ping to test a network connection.

12.10 HANDS-ON ACTIVITY 10: SIMPLE NETWORK SET UP USING A DAISY CHAIN

Parts and Equipment

- Two computers
- Two hubs
- Three straight-through CAT-5 UTP cables with RJ-45 connectors.

Procedure

1. Find the IP address of the computers (see Hands-On Activity 2).
2. Connect hubs together and connect the first hub to the first computer and the second hub to the second computer using CAT-5 cable by inserting the RJ-45 connectors to the proper ports on each device (see Figure 12–10).
3. Make sure the hubs and computers are compatible with each other in terms of speed.
4. Click the Windows logo in the lower left corner of the screen to open the Start Menu. Type **cmd** in the search box to open a Command Prompt window.
5. At the command prompt, C:\, type ping **192. 168.*xyz.a*** (the IP address of the second computer) and press Enter. You will either receive four "receiving" replies, or if the set up of the network has dnot been done property, you will see an error message (the error light on the hubs may also blink or turn red).

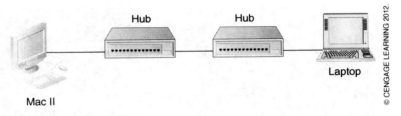

Figure 12–10: A simple network setup using a daisy chain

Appendix A: ASCII Code

ASCII	Hexdecimal	Octal	Character	Description
0	0	0		null
1	1	1		start of heading
2	2	2		start of text
3	3	3		end of text
4	4	4		end of transmission
5	5	5		enquiry
6	6	6		acknowledge
7	7	7		bell
8	8	10		backspace
9	9	11		horizontal tab
10	A	12		new line
11	B	13		vertical tab
12	C	14		new page
13	D	15		carriage return
14	E	16		shift out
15	F	17		shift in
16	10	20		data link escape
17	11	21		device control 1
18	12	22		device control 2
19	13	23		device control 3
20	14	24		device control 4

American Standard Code for Information Interchange (ASCII Code) (*Continued*)

ASCII	Hexdecimal	Octal	Character	Description
21	15	25		negative acknowledge
22	16	26		synchronous idle
23	17	27		end of trans. block
24	18	30		cancel
25	19	31		end of medium
26	1A	32		substitute
27	1B	33		escape
28	1C	34		file separator
29	1D	35		group separator
30	1E	36		record separator
31	1F	37		unit separator
32	20	40		space
33	21	41	!	
34	22	42	"	
35	23	43	#	
36	24	44	$	
37	25	45	%	
38	26	46	&	
39	27	47	'	
40	28	50	(	
41	29	51	)	
42	2A	52	*	
43	2B	53	+	
44	2C	54	,	
45	2D	55	-	
46	2E	56	.	
47	2F	57	/	
48	30	60	0	
49	31	61	1	
50	32	62	2	
51	33	63	3	
52	34	64	4	
53	35	65	5	
54	36	66	6	
55	37	67	7	

American Standard Code for Information Interchange (ASCII Code) (*Continued*)

ASCII	Hexdecimal	Octal	Character	Description
56	38	70	8	
57	39	71	9	
58	3A	72	:	
59	3B	73	;	
60	3C	74	<	
61	3D	75	=	
62	3E	76	>	
63	3F	77	?	
64	40	100	@	
65	41	101	A	
66	42	102	B	
67	43	103	C	
68	44	104	D	
69	45	105	E	
70	46	106	F	
71	47	107	G	
72	48	110	H	
73	49	111	I	
74	4A	112	J	
75	4B	113	K	
76	4C	114	L	
77	4D	115	M	
78	4E	116	N	
79	4F	117	O	
80	50	120	P	
81	51	121	Q	
82	52	122	R	
83	53	123	S	
84	54	124	T	
85	55	125	U	
86	56	126	V	
87	57	127	W	
88	58	130	X	
89	59	131	Y	
90	5A	132	Z	
91	5B	133	[	

American Standard Code for Information Interchange (ASCII Code) (*Continued*)

ASCII	Hexdecimal	Octal	Character	Description
92	5C	134	\	
93	5D	135	]	
94	5E	136	^	
95	5F	137	_	
96	60	140	`	
97	61	141	a	
98	62	142	b	
99	63	143	c	
100	64	144	d	
101	65	145	e	
102	66	146	f	
103	67	147	g	
104	68	150	h	
105	69	151	i	
106	6A	152	j	
107	6B	153	k	
108	6C	154	l	
109	6D	155	m	
110	6E	156	n	
111	6F	157	o	
112	70	160	p	
113	71	161	q	
114	72	162	r	
115	73	163	s	
116	74	164	t	
117	75	165	u	
118	76	166	v	
119	77	167	w	
120	78	170	x	
121	79	171	y	
122	7A	172	z	
123	7B	173	{	
124	7C	174			
125	7D	175	}	
126	7E	176	~	
127	7F	177	DEL	

American Standard Code for Information Interchange (ASCII Code)

Decimal	Hexadecimal	Symbol	Description
0	00	NUL	Null char
1	01	SOH	Start of Heading
2	02	STX	Start of Text
3	03	ETX	End of Text
4	04	EOT	End of Transmission
5	05	ENQ	Enquiry
6	06	ACK	Acknowledgment
7	07	BEL	Bell
8	08	BS	Back Space
9	09	HT	Horizontal Tab
10	0A	LF	Line Feed
11	0B	VT	Vertical Tab
12	0C	FF	Form Feed
13	0D	CR	Carriage Return
14	0E	SO	Shift Out / X-On
15	0F	SI	Shift In / X-Off
16	10	DLE	Data Line Escape
17	11	DC1	Device Control 1 (oft. XON)
18	12	DC2	Device Control 2
19	13	DC3	Device Control 3 (oft. XOFF)
20	14	DC4	Device Control 4
21	15	NAK	Negative Acknowledgement
22	16	SYN	Synchronous Idle
23	17	ETB	End of Transmit Block
24	18	CAN	Cancel
25	19	EM	End of Medium
26	1A	SUB	Substitute
27	1B	ESC	Escape
28	1C	FS	File Separator
29	1D	GS	Group Separator
30	1E	RS	Record Separator
31	1F	US	Unit Separator
32	20		Space
33	21	!	Exclamation mark
34	22	"	Double quotes (or speech marks)
35	23	#	Number

Extended ASCII Codes (*Continued*)

Decimal	Hexadecimal	Symbol	Description
36	24	$	Dollar
37	25	%	Procenttecken
38	26	&	Ampersand
39	27	'	Single quote
40	28	(	Open parenthesis (or open bracket)
41	29	)	Close parenthesis (or close bracket)
42	2A	*	Asterisk
43	2B	+	Plus
44	2C	,	Comma
45	2D	-	Hyphen
46	2E	.	Period, dot or full stop
47	2F	/	Slash or divide
48	30	0	Zero
49	31	1	One
50	32	2	Two
51	33	3	Three
52	34	4	Four
53	35	5	Five
54	36	6	Six
55	37	7	Seven
56	38	8	Eight
57	39	9	Nine
58	3A	:	Colon
59	3B	;	Semicolon
60	3C	<	Less than (or open angled bracket)
61	3D	=	Equals
62	3E	>	Greater than (or close angled bracket)
63	3F	?	Question mark
64	40	@	At symbol
65	41	A	Uppercase A
66	42	B	Uppercase B
67	43	C	Uppercase C
68	44	D	Uppercase D
69	45	E	Uppercase E
70	46	F	Uppercase F
71	47	G	Uppercase G

Extended ASCII Codes (*Continued*)

Decimal	Hexadecimal	Symbol	Description
72	48	H	Uppercase H
73	49	I	Uppercase I
74	4A	J	Uppercase J
75	4B	K	Uppercase K
76	4C	L	Uppercase L
77	4D	M	Uppercase M
78	4E	N	Uppercase N
79	4F	O	Uppercase O
80	50	P	Uppercase P
81	51	Q	Uppercase Q
82	52	R	Uppercase R
83	53	S	Uppercase S
84	54	T	Uppercase T
85	55	U	Uppercase U
86	56	V	Uppercase V
87	57	W	Uppercase W
88	58	X	Uppercase X
89	59	Y	Uppercase Y
90	5A	Z	Uppercase Z
91	5B	[	Opening bracket
92	5C	\	Backslash
93	5D	]	Closing bracket
94	5E	^	Caret - circumflex
95	5F	_	Underscore
96	60	`	Grave accent
97	61	a	Lowercase a
98	62	b	Lowercase b
99	63	c	Lowercase c
100	64	d	Lowercase d
101	65	e	Lowercase e
102	66	f	Lowercase f
103	67	g	Lowercase g
104	68	h	Lowercase h
105	69	i	Lowercase i
106	6A	j	Lowercase j
107	6B	k	Lowercase k

Extended ASCII Codes (*Continued*)

Decimal	Hexadecimal	Symbol	Description
108	6C	l	Lowercase l
109	6D	m	Lowercase m
110	6E	n	Lowercase n
111	6F	o	Lowercase o
112	70	p	Lowercase p
113	71	q	Lowercase q
114	72	r	Lowercase r
115	73	s	Lowercase s
116	74	t	Lowercase t
117	75	u	Lowercase u
118	76	v	Lowercase v
119	77	w	Lowercase w
120	78	x	Lowercase x
121	79	y	Lowercase y
122	7A	z	Lowercase z
123	7B	{	Opening brace
124	7C	|	Vertical bar
125	7D	}	Closing brace
126	7E	~	Equivalency sign - tilde
127	7F		Delete
128	80	€	Euro sign
129	81		
130	82	‚	Single low-9 quotation mark
131	83	ƒ	Latin small letter f with hook
132	84	„	Double low-9 quotation mark
133	85	…	Horizontal ellipsis
134	86	†	Dagger
135	87	‡	Double dagger
136	88	ˆ	Modifier letter circumflex accent
137	89	‰	Per mille sign
138	8A	Š	Latin capital letter S with caron
139	8B	‹	Single left-pointing angle quotation
140	8C	Œ	Latin capital ligature OE
141	8D		
142	8E	Ž	Latin capital letter Z with caron
143	8F		

Extended ASCII Codes (*Continued*)

Decimal	Hexadecimal	Symbol	Description
144	90		
145	91	'	Left single quotation mark
146	92	'	Right single quotation mark
147	93	"	Left double quotation mark
148	94	"	Right double quotation mark
149	95	•	Bullet
150	96	–	En dash
151	97	—	Em dash
152	98	~	Small tilde
153	99	™	Trade mark sign
154	9A	š	Latin small letter S with caron
155	9B	›	Single right-pointing angle quotation mark
156	9C	œ	Latin small ligature oe
157	9D		
158	9E	ž	Latin small letter z with caron
159	9F	Ÿ	Latin capital letter Y with dieresis or dieresis
160	A0		Non-breaking space
161	A1	¡	Inverted exclamation mark
162	A2	¢	Cent sign
163	A3	£	Pound sign
164	A4	¤	Currency sign
165	A5	¥	Yen sign
166	A6	¦	Pipe, Broken vertical bar
167	A7	§	Section sign
168	A8	¨	Spacing dieresis - umlaut
169	A9	©	Copyright sign
170	AA	ª	Feminine ordinal indicator
171	AB	«	Left double angle quotes
172	AC	¬	Not sign
173	AD		Soft hyphen
174	AE	®	Registered trade mark sign
175	AF	¯	Spacing macron – overline
176	B0	°	Degree sign
177	B1	±	Plus-or-minus sign
178	B2	²	Superscript two - squared
179	B3	³	Superscript three - cubed

Extended ASCII Codes (*Continued*)

Decimal	Hexadecimal	Symbol	Description
180	B4	´	Acute accent - spacing acute
181	B5	µ	Micro sign
182	B6	¶	paragraph sign
183	B7	·	Middle dot - Georgian comma
184	B8	¸	Spacing cedilla
185	B9	¹	Superscript one
186	BA	º	Masculine ordinal indicator
187	BB	»	Right double angle quotes
188	BC	¼	Fraction one quarter
189	BD	½	Fraction one half
190	BE	¾	Fraction three quarters
191	BF	¿	Inverted question mark
192	C0	À	Latin capital letter A with grave
193	C1	Á	Latin capital letter A with acute
194	C2	Â	Latin capital letter A with circumflex
195	C3	Ã	Latin capital letter A with tilde
196	C4	Ä	Latin capital letter A with dieresis
197	C5	Å	Latin capital letter A with ring above
198	C6	Æ	Latin capital letter AE
199	C7	Ç	Latin capital letter C with cedilla
200	C8	È	Latin capital letter E with grave
201	C9	É	Latin capital letter E with acute
202	CA	Ê	Latin capital letter E with circumflex
203	CB	Ë	Latin capital letter E with dieresis
204	CC	Ì	Latin capital letter I with grave
205	CD	Í	Latin capital letter I with acute
206	CE	Î	Latin capital letter I with circumflex
207	CF	Ï	Latin capital letter I with dieresis
208	D0	Ð	Latin capital letter ETH
209	D1	Ñ	Latin capital letter N with tilde
210	D2	Ò	Latin capital letter O with grave
211	D3	Ó	Latin capital letter O with acute
212	D4	Ô	Latin capital letter O with circumflex
213	D5	Õ	Latin capital letter O with tilde
214	D6	Ö	Latin capital letter O with dieresis

Extended ASCII Codes (*Continued*)

Decimal	Hexadecimal	Symbol	Description
215	D7	×	Multiplication sign
216	D8	Ø	Latin capital letter O with slash
217	D9	Ù	Latin capital letter U with grave
218	DA	Ú	Latin capital letter U with acute
219	DB	Û	Latin capital letter U with circumflex
220	DC	Ü	Latin capital letter U with dieresis
221	DD	Ý	Latin capital letter Y with acute
222	DE	Þ	Latin capital letter THORN
223	DF	ß	Latin small letter sharp s - ess-zed
224	E0	à	Latin small letter a with grave
225	E1	á	Latin small letter a with acute
226	E2	â	Latin small letter a with circumflex
227	E3	ã	Latin small letter a with tilde
228	E4	ä	Latin small letter a with dieresis
229	E5	å	Latin small letter a with ring above
230	E6	æ	Latin small letter ae
231	E7	ç	Latin small letter c with cedilla
232	E8	è	Latin small letter e with grave
233	E9	é	Latin small letter e with acute
234	EA	ê	Latin small letter e with circumflex
235	EB	ë	Latin small letter e with dieresis
236	EC	ì	Latin small letter i with grave
237	ED	í	Latin small letter i with acute
238	EE	î	Latin small letter i with circumflex
239	EF	ï	Latin small letter i with dieresis
240	F0	ð	Latin small letter eth
241	F1	ñ	Latin small letter n with tilde
242	F2	ò	Latin small letter o with grave
243	F3	ó	Latin small letter o with acute
244	F4	ô	Latin small letter o with circumflex
245	F5	õ	Latin small letter o with tilde
246	F6	ö	Latin small letter o with dieresis
247	F7	÷	Division sign
248	F8	ø	Latin small letter o with slash
249	F9	11111001	Latin small letter u with grave

Extended ASCII Codes (*Continued*)

Decimal	Hexadecimal	Symbol	Description
250	FA	ú	Latin small letter u with acute
251	FB	û	Latin small letter u with circumflex
252	FC	ü	Latin small letter u with dieresis
253	FD	ý	Latin small letter y with acute
254	FE	þ	Latin small letter thorn
255	FF	ÿ	Latin small letter y with dieresis

Extended ASCII Codes

Hexadecimal	6-Ternary code	Hexadecimal	6-Ternary code	Hexadecimal	6-Ternary code
00	+ − 0 0 + −	10	+ 0 + − − 0	20	0 0 − + + −
01	0 + − + − 0	11	+ + 0 − 0 −	21	− − + 0 0 +
02	+ − 0 + − 0	12	+ 0 + − 0 −	22	+ + − 0 + −
03	− 0 + + − 0	13	0 + + − 0 −	23	+ + − 0 − +
04	− 0 + 0 + −	14	0 + + − − 0	24	0 0 + 0 − +
05	0 + − − 0 +	15	+ + 0 0 − −	25	0 0 + 0 + −
06	+ − 0 − 0 +	16	+ 0 + 0 − −	26	0 0 − 0 0 +
07	− 0 + − 0 +	17	0 + + 0 − −	27	− − + + + −
08	− + 0 0 + −	18	0 + − 0 + −	28	− 0 − + + 0
09	0 − + + − 0	19	0 + − 0 − +	29	− − 0 + 0 +
0A	− + 0 + − 0	1A	0 + − + + −	2A	− 0 − + 0 +
0B	+ 0 − + − 0	1B	0 + − 0 0 +	2B	0 − − + 0+
0C	+ 0 − 0 + −	1C	0 − + 0 0 +	2C	0 − − + + 0
0D	0 − + − 0 +	1D	0 − + + + −	2D	− − 0 0 + +
0E	− + 0 − 0 +	1E	0 − + 0 − +	2E	− 0 − 0 + +
0F	+ 0 − − 0 +	1F	0 − + 0 + −	2F	0 − − 0 ++

Hexadecimal	6-Ternary code	Hexadecimal	6-Ternary code	Hexadecimal	6-Ternary code
30	+ − 0 0 − +	40	+ 0 + 0 0 −	50	+ 0 + − − +
31	0 + − − + 0	41	+ + 0 0 − 0	51	+ + 0 − + −
32	+ − 0 − + 0	42	+ 0 + 0 − 0	52	+ 0 + − + −
33	− 0 + − + 0	43	0 + + 0 − 0	53	0 + + − + −
34	− 0 + 0 − +	44	0 + + 0 0 −	54	0 + + − − +
35	0 + − + 0 −	45	+ + 0 − 0 0	55	+ + 0 + − −
36	+ − 0 + 0 −	46	+ 0 + − 0 0	56	+ 0 + + − −
37	− 0 + + 0 −	47	0 + + − 0 0	57	0 + + + − −
38	− + 0 0 − +	48	0 0 0 + 0 0	58	+ + + 0 − −
39	0 − + − + 0	49	0 0 0 − + +	59	+ + + − 0 −
3A	− + 0 − + 0	4A	0 0 0 + − +	5A	+ + + − − 0
3B	+ 0 − − + 0	4B	0 0 0 + + −	5B	+ + 0 − − 0
3C	+ 0 − 0 − +	4C	0 0 0 + 0 0	5C	+ + 0 − − +
3D	0 − + + 0 −	4D	0 0 0 − + +	5D	+ + 0 0 0 −
3E	− + 0 + 0 −	4E	0 0 0 + − +	5E	− − + + + 0
3F	+ 0 − + 0 −	4F	0 0 0 + + −	5F	0 0 − + + 0

Hexadecimal	6-Ternary code	Hexadecimal	6-Ternary code	Hexadecimal	6-Ternary code
60	0 − 0 + + 0	70	− + + 0 0 0	80	+ − + 0 0 −
61	0 0 − + 0 +	71	+ − + 0 0 0	81	+ + − 0 − 0
62	0 − 0 + 0 +	72	+ + − 0 0 0	82	+ − + 0 − 0
63	− 0 0 + 0 +	73	0 0 + 0 0 0	83	− + + 0 − 0
64	− 0 0 + + 0	74	− 0 + 0 0 0	84	− + + 0 0 −
65	0 0 − 0 + +	75	0 − + 0 0 0	85	+ + − − 0 0
66	0 − 0 0 + +	76	+ 0 − 0 0 0	86	+ − + − 0 0
67	− 0 0 0 + +	77	0 + − 0 0 0	87	− + + − 0 0
68	− + − + + 0	78	0 − − + + +	88	0 + 0 0 0 −
69	− − + + 0 +	79	− 0 − + + +	89	0 0 + 0 − 0
6A	− + − + 0 +	7A	− − 0 + + +	8A	0 + 0 0 − 0
6B	+ − − + 0 +	7B	− − 0 + + −	8B	+ 0 0 0 − 0
6C	− + − + + 0	7C	+ + − 0 0 −	8C	+ 0 0 0 0 −
6D	− − + + 0 +	7D	0 0 + 0 0 −	8D	0 0 + − 0 0
6E	− + − + 0 +	7E	+ + − − − +	8E	0 + 0 − 0 0
6F	+ − − + 0 +	7F	0 0 + − − +	8F	+ 0 0 − 0 0

Hexadecimal	6-Ternary code	Hexadecimal	6-Ternary code	Hexadecimal	6-Ternary code
90	+ − + − − +	A0	0 + 0 + − +	B0	0 − 0 0 0 +
91	+ + − − + −	A1	0 0 − + − +	B1	0 0 − 0 + 0
92	+ − + − − +	A2	0 − 0 + − +	B2	0 − 0 0 + 0
93	+ + − − + −	A3	− 0 0 + − +	B3	− 0 0 0 + 0
94	− + + − − +	A4	− 0 0 + + −	B4	− 0 0 0 0 +
95	+ + − + − −	A5	0 0 − − + +	B5	0 0 − + 0 0
96	+ − + + − −	A6	0 − 0 − + +	B6	0 − 0 + 0 0
97	− + + + − −	A7	− 0 0 − + +	B7	− 0 0 + 0 0
98	0 + 0 − − +	A8	− + − + + −	B8	− + − 0 0 +
99	0 0 + − + −	A9	− − + + − +	B9	− − + 0 + 0
9A	0 + 0 − + −	AA	− + − + − +	BA	− + − 0 + 0
9B	+ 0 0 − + −	AB	+ − − + − +	BB	+ − − 0 + 0
9C	+ 0 0 − − +	AC	+ − − + + −	BC	+ − − 0 0 +
9D	0 0 + + − −	AD	− − + − + +	BD	− − + + 0 0
9E	0 + 0 + − −	AE	− + − − + +	BE	− + − + 0 0
9F	+ 0 0 + − −	AF	+ − − − + +	BF	+ − − + 0 0

Hexadecimal	6-Ternary code	Hexadecimal	6-Ternary code	Hexadecimal	6-Ternary code
C0	+ − + 0 + −	D0	+ − + 0 − +	E0	+ − 0 0 − +
C1	+ + − + − 0	D1	+ + − − + 0	E1	0 + − + − +
C2	+ − + + − 0	D2	+ − + − + 0	E2	+ − 0 + − +
C3	− + + + − 0	D3	− + + − + 0	E3	− 0 + + − +
C4	− + + 0 + −	D4	− + + 0 − +	E4	− 0 + + + −
C5	+ + − − 0 +	D5	+ + − + 0 −	E5	0 + − − +0
C6	+ − + − 0 +	D6	+ − + + 0 −	E6	+ − 0 − + +
C7	− + + − 0 +	D7	− + + + 0 −	E7	− 0 + − + +
C8	0 + 0 0 + −	D8	0 + 0 0 − +	E8	− + 0 + + −
C9	0 0 + + − 0	D9	0 0 + − + 0	E9	0 − + + − +
CA	0 + 0 + − 0	DA	0 + 0 − + 0	EA	− + 0 + − +
CB	+ 0 0 + − 0	DB	+ 0 0 − + 0	EB	+ 0 − + − +
CC	+ 0 0 0 + −	DC	+ 0 0 0 − +	EC	+ 0 − + + −
CD	0 0 + − 0 +	DD	0 0 + + 0 −	ED	0 − + − + +
CE	0 + 0 − 0 +	DE	0 + 0 + 0 −	EE	− + 0 − + +
CF	+ 0 0 − 0 +	DF	+ 0 0 + 0 −	EF	+ 0 − − + +

Hexadecimal	6-Ternary code
F0	+ − 0 0 0 +
F1	0 + − 0 + 0
F2	+ − 0 0 + 0
F3	− 0 + 0 + 0
F4	− 0 + 0 0 +
F5	0 + − + 0 0
F6	+ − 0 + 0 0
F7	− 0 + + 0 0
F8	− + 0 0 0 +
F9	0 − + 0 + 0
FA	− + − + 0 +
FB	+ 0 − 0 + 0
FC	+ 0 − 0 0 +
FD	0 − + + 0 0
FE	− + 0 + 0 0
FF	+ 0 − + 0 0

Index

CPSIA information can be obtained
at www.ICGtesting.com
Printed in the USA
FFOW03n1315101215
19458FF